Public versus Private Transport

C. A. NASH

British Rail Lecturer in Rail Transport,
Institute for Transport Studies, Leeds University

M

First published 1976 by
THE MACMILLAN PRESS LTD
London and Basingstoke
Associated companies in New York Dublin
Melbourne Johannesburg and Madras

SBN 333 15405 3

Printed in Great Britain by
THE ANCHOR PRESS LTD
Tiptree, Essex

Editors' Preface to Macmillan Studies in Economics

The rapid growth of academic literature in the field of economics has posed serious problems for both students and teachers of the subject. The latter find it difficult to keep pace with more than a few areas of the subject so that an inevitable trend towards specialism emerges. The student quickly loses perspective as the maze of theories and models grows, particularly at a time when so much reappraisal of the established paradigms is taking place.

The aim of the 'Macmillan Studies in Economics' is to offer students, and perhaps some teachers as well, short, reasonably critical overviews of developments in selected areas of economics, particularly those in which current controversies are to be found. As far as possible the titles have been selected to form an integrated whole, although inevitably whole areas have been neglected as being unsuited to the style, format and length of the titles in the series.

In some cases the volumes are rather more like essays than surveys. In most cases, however, the aim is to survey the salient literature in a critical fashion. The level of understanding required to read the volumes varies with the complexity of the subject, but they have been generally written to suit the second- and third-year undergraduate seeking to place his reading of the detailed literature in an over-all context. They are *not* textbooks. Instead they seek to give the kind of perspective that might well be lost by reading longer textbooks on their own, or by reading articles in journals. In particular, they should be most suited to pre-examination revision periods. They are not intended to substitute for the essential reading and assimilation of the original works that they seek to survey and assess.

MACMILLAN STUDIES IN ECONOMICS

General Editors: D. C. ROWAN and G. R. FISHER

Executive Editor: D. W. PEARCE

Published

R. W. Anderson: THE ECONOMICS OF CRIME
John Burton: WAGE INFLATION
Ben Fine: MARX'S 'CAPITAL'
Douglas Fisher: MONETARY POLICY
Miles Fleming: MONETARY THEORY
C. J. Hawkins and D. W. Pearce: CAPITAL INVESTMENT APPRAISAL
C. J. Hawkins: THEORY OF THE FIRM
David F. Heathfield: PRODUCTION FUNCTIONS
Dudley Jackson: POVERTY
P. N. Junankar: INVESTMENT: THEORIES AND EVIDENCE
J. E. King: LABOUR ECONOMICS
John King and Philip Regan: RELATIVE INCOME SHARES
J. A. Kregel: THE THEORY OF ECONOMIC GROWTH
J. A. Kregel: THEORY OF CAPITAL
Richard Lecomber: ECONOMIC GROWTH VERSUS THE ENVIRONMENT
George McKenzie: THE MONETARY THEORY OF INTERNATIONAL TRADE
David J. Mayston: THE IDEA OF SOCIAL CHOICE
C. A. Nash: PUBLIC VERSUS PRIVATE TRANSPORT
S. K. Nath: A PERSPECTIVE OF WELFARE ECONOMICS
Anthony Peaker: ECONOMIC GROWTH IN MODERN BRITAIN
D. W. Pearce: COST-BENEFIT ANALYSIS
Maurice Peston: PUBLIC GOODS AND THE PUBLIC SECTOR
Nicholas Rau: TRADE CYCLES: THEORY AND EVIDENCE
David Robertson: INTERNATIONAL TRADE POLICY
Charles K. Rowley: ANTITRUST AND ECONOMIC EFFICIENCY
C. H. Sharp: TRANSPORT ECONOMICS
G. K. Shaw: FISCAL POLICY
R. Shone: THE PURE THEORY OF INTERNATIONAL TRADE
M. J. Stabler: AGRICULTURAL ECONOMICS AND RURAL LAND-USE
Frank J. B. Stilwell: REGIONAL ECONOMIC POLICY
A. P. Thirlwall: FINANCING ECONOMIC DEVELOPMENT
John Vaizey: THE ECONOMICS OF EDUCATION
J. van Doorn: DISEQUILIBRIUM ECONOMICS
Peter A. Victor: ECONOMICS OF POLLUTION
Grahame Walshe: INTERNATIONAL MONETARY REFORM
Michael G. Webb: PRICING POLICIES FOR PUBLIC ENTERPRISES
E. Roy Weintraub: CONFLICT AND CO-OPERATION IN ECONOMICS
E. Roy Weintraub: GENERAL EQUILIBRIUM THEORY

Forthcoming

S. Charles HOUSING ECONOMICS
C. Collis and K. Turner: ECONOMICS OF PLANNING
B. Morgan: MONETARISM AND KEYNESIANISM
A. Ziderman: MANPOWER TRAINING: THEORY AND POLICY

Contents

Acknowledgements

This book was written largely whilst I was in the post of Lecturer in Economics at Southampton University, and I should like to thank colleagues there, in particular David Pearce, for all they taught me about the debate on the environment. I have benefited greatly by receiving valuable and extensive comments from David Hensher and John Stanley (Commonwealth Bureau of Roads, Australia), George Docwra, (University of Queensland), Tony Jennings (University of Leicester), and Neill Schofield (Department of the Environment), all of whom took great trouble to read my first confused thoughts on this subject. Needless to say responsibility for the final version is solely my own, and the views contained in it are not necessarily those of any other person or organisation. Finally, I should like to thank Ms A. Bartoli of Southampton University who typed the first draft, and my wife, Claire, who typed the final version – and corrected the grammar!

1 The Public- versus Private-Transport Debate

> For a fraction of the money the nation is currently spending on the maintenance of private cars and on the Government services necessary to keep the traffic moving . . . we could provide a comfortable, frequent and highly efficient public transport service.
>
> (Mishan [1] p. 128)

> Since 1945 it has been the policy of successive British Governments to subsidise rail transport and to tax road transport. . . . It is difficult to justify this policy on economic grounds.
>
> (Roth [2] p. 22)

The debate between the proponents of public and private transport has passed through two main phases. In the first, the argument was that, simply in terms of the capital and operating costs of the transport system, the institutional arrangements gave too much encouragement to private transport as opposed to public. Originally the debate was seen largely as one concerning the correct recovery of the 'track costs' incurred in the provision of roads, in order that road transport should not benefit from the fact that the state provided its track whereas railway companies were responsible for the provision of their own track. This argument was influential in the controversy which led to the introduction of licensing systems for the operation of public road goods and passenger services in the 1930s.[1] However, at that time, the rivalry was seen largely as one

[1] For a description of these licensing systems and their effects see [3]. The other main argument put forward in favour of licensing was that excessive competition was leading to instability and poor safety standards in road transport.

between *public* road and *public* rail services. Thus licensing did nothing to prevent the spectacular rise of *private* transport to a situation where it accounted for 80 per cent of passenger-miles and 26 per cent of freight ton-miles in 1972.[1] Indeed, by hampering the growth of public road services, the licensing system may actually have encouraged the trend towards private transport.

The second stage of the debate has been concerned with the external effects of transport systems. Even at the time of the Royal Commission on Transport of 1930 [4], there was concern about the rise in road congestion and accidents. Since then the debate has widened to include questions of noise, air and visual pollution (indeed, the effects of motor vehicles on the environment in general), patterns of land-use, and the depletion of stocks of non-renewable resources (especially oil). Whereas for passenger transport, public transport, whether by road or rail, yields very different effects from private transport in terms of these variables, for freight transport it is choice of mode which is more important.

We have already indicated the dominant role played by private (road) transport in the passenger sector. When public and private road transport are added together, a similar dominance appears in the freight sector; as a whole, roads catered for 92 per cent of passenger-miles and 64 per cent of freight ton-miles in the United Kingdom in 1972.[2] In itself this does not necessarily prove the superiority of road transport. In the first place, the prices charged to users of different modes of transport may not reflect accurately the costs incurred in providing those facilities. Secondly, transport users may misperceive these costs, may be unaware of the range of alternatives available, or may allocate traffic on the basis of some criterion which fails to pay adequate attention to the alterna-

[1] By private freight transport, we refer to carriage on own-account rather than for hire and reward. These estimates are given in the *Annual Abstract of Statistics, 1973* (London: H.M.S.O., 1973) tables 236, 245.

[2] Ibid. It should be noted that taking ton-miles is not necessarily the best method of aggregating that highly diverse range of outputs, that is freight transport. There is no particular reason to give more weight to heavy items than to light ones of equivalent or greater bulk. If prices paid for transport services are used as weights in calculating relative output, the dominance of road transport is even more pronounced. See [5].

tives. For instance, it may be organisationally simplest to use private transport automatically rather than to seek the best mode for each traffic, but that does not imply that the information and decision costs of the latter course of action necessarily outweigh its benefits, even to the firm itself.

Chapters 2 and 3 of this book review the current state of the controversy over the measurement of the social costs of different modes of transport, including track costs, costs of congestion and environmental effects. Chapter 4 asks whether the price mechanism can be adapted adequately to reflect all these extra factors. It is concluded that, generally, it cannot; and that a wide variety of tools of planning, traffic management and administrative controls will be needed to improve resource allocation in the transport sector. Subsequent chapters deal with the desirability and methods of influencing modal choice in different sectors of the transport market. For convenience these have been grouped into urban passenger transport, inter-urban and rural-passenger transport, and freight transport. Most of the examples are taken from U.K. experience, although the methodology is more widely applicable. Finally, in Chapter 8, we present our conclusions on the private- versus public-transport debate. In brief, these are that there is a strong case for promoting public rather than private transport in urban areas and on heavily used recreational routes. In the case of freight transport it does appear that more use should be made of public rail transport for trunk haulage, and public road transport (operating from peripheral break-bulk depots) for urban delivery work. These reforms need not impose high costs on transport users and consumers. For most other transport, diversion from private to public transport would impose high cost, in the absence of major changes in residential and commercial location and density.

This is a book about choice of transport policy; an issue which – as with all policy matters – has its roots in the ethical beliefs of those doing the choosing. That transport is primarily desired as an intermediate good,[1] for which an apparently

[1] Even leisure travel by individuals is usually an input into the consumption of some other good, for example sport, cinema, visits to places or people, although travel is sometimes desired as a good in its own right.

objective criterion – that of minimising the 'true' or social costs of an 'adequate' level of provision of services – is often put forward, is apt to confuse the reader. The question of what is and what is not a social cost, and of how these costs are to be measured, is an ethical issue, as is the measurement of the benefits of different levels of provision of transport services to determine their 'adequacy'. For instance, use of unweighted market prices or consumers' and producers' surplus measures, as is usual in current cost–benefit analysis of transport systems,[1] implies that the current distribution of income is regarded as optimal [6]. If this is not the case, then the question of changing modal split in transport raises distributional issues, requiring the tracing of the ultimate incidence of the costs of such policies, which may be far removed from the immediate transport user.

Corresponding with the variety of value judgements on how to measure costs and benefits of transport services, and with the variety of forecasts of future technological and social trends, many possible transport policies emerge. It is hoped that, even if the reader does not share the views of the author on these subjects, this book may help him (or her) to understand the issues more clearly, and that the methodology may help him to formulate his own conclusions on the issue of public and private transport.

[1] The technique of cost–benefit analysis, and the application of the principles of welfare economics to problems of pricing and investment policy in the transport sector, are the subject of companion volumes in this series by D. W. Pearce [7] and C. H. Sharp [8] and are not considered in detail here.

2 Track Costs

The appropriateness of the current method of charging for transport infrastructure is still often seen as a key issue in transport-mode-choice decisions. The price paid for transport services by the private or public road-transport operator in the United Kingdom comprises partly his money outlay on fuel and on other inputs bought in the market, partly the opportunity cost of other resources (for example, his own time) used in providing the services, and partly taxes (fuel tax, licence duty and – in the case of private individuals – value-added tax) paid to the state. He pays no direct toll for the use of road space, which is provided by the state. Yet this has not stopped commentators from comparing the taxes raised from different sectors of the market with the track costs which these impose. For instance, the British Railways Board has argued that heavy lorries are responsible for over 70 per cent of the costs of constructing new roads, and are currently cross-subsidised by other road users [9]. On the other hand, the Transport Holding Company has argued that the use of roads confers external benefits on the community, which should be allowed for in setting prices ([10] annex 3).

The most important official document on the track-costs debate was published by the Department of the Environment in 1968 [10]. This compared the revenue raised from four separate categories of road vehicle – cars, light vans, public-service vehicles, and heavy-goods vehicles – with the costs incurred in providing for each. It was found that, whilst each category more than covered its costs, the ratio of revenue to costs was lower for heavy-goods vehicles than for the other categories (with the exception of public-service vehicles, which were considered to be a 'special case'), and this was used to justify proposals for an extra road-haulage charge on heavy-goods vehicles.

Unfortunately, the study ran into the usual joint-cost problems associated with the provision of transport facilities. First, there is the question of the allocation of maintenance costs. Surface maintenance costs were largely allocated to heavy-goods vehicles, although not to as great an extent as would be implied by the controversial 'fourth power' rule.[1] Remaining maintenance costs were allocated in proportion to the mileage run by each class of vehicle.

Second, there are those costs incurred in cleansing, lighting, policing and administering the road system. Whilst policing costs may vary to some extent with the level and composition of traffic flow, for the most part these appear to be classic cases of overhead joint costs, Nevertheless, they were allocated amongst users in proportion to vehicle-miles run.

However, the greatest problems were produced by an item much bigger than these categories of cost – the capital cost of new construction and major improvements. Two treatments of this item were proposed. The first was to charge interest on the capital value of existing roads. This approach foundered on the problems of measuring the replacement (as opposed to the historic) cost of roads. The alternative that was finally adopted was to treat capital expenditure on roads as current expenditure, to be recovered in the year in which it was incurred.[2] Of this expenditure, 17·5 per cent was regarded as being incurred exclusively on behalf of heavy-goods vehicles. The remainder was allocated on the basis of vehicle-miles, except that heavy vehicles were given twice the weighting of light ones.

The arbitrary nature of many of these cost allocations scarcely needs further comment. What does call for further consideration is the principle on which this allocation is based – that each group of users of a given road should exactly pay the costs for which it is responsible. For this implies charging each group of users the average cost which it imposes, and will only coincide with the usual prescription for an efficient allocation

[1] That relative damage to the road surface rises in proportion to the fourth power of the increase in laden axle weight.

[2] The method of treating capital costs was criticised by Munby [11], leading subsequently to controversy in the *Journal of Transport Economics and Policy*, vol. 2 no. 3 (1968).

of resources – marginal-cost pricing – if there are constant returns to scale. Even this result holds only if all costs are valued in replacement terms; in terms of historic cost the sector is likely to show a profit.

When they refer to resource-allocation problems, the authors are quite clear that what they wish to measure is long-run marginal cost [10] – hence the inclusion of overheads and capital costs and the omission of congestion costs. But long-run marginal cost is only an appropriate basis for pricing policy if it is believed either that capacity is optimal, or that it will be optimally adjusted within a sufficiently short space of time to make the expense of charging short-run marginal cost in the interim, and then changing price again, not worthwhile [12].

Consider an individual road. As the traffic flow on it increases, so congestion causes speed to drop. This results in an upward-sloping marginal private cost curve (*MPC*) as in Figure 1. Briefly the explanation for this is partly that congestion tends to raise vehicle operating costs, but more significantly that one of the costs of making a journey is regarded as the opportunity cost of the time expended on it, and as speeds fall journey times increase.[1] Above this is shown the marginal social cost curve (*MSC*). This is derived from the *MPC* schedule by adding to it increased road maintenance costs, and the extra costs imposed on other vehicles by the increased congestion resulting from the presence of one extra vehicle (we postpone consideration of other variations between *MPC* and *MSC* until Chapter 3). The standard argument is that the optimal level of user taxation is given by t in Figure 1, being that level leading to a traffic flow at which the willingness to pay of the motorist for the marginal journey equals the marginal social cost which he imposes.[2]

Now let us assume that the road-building authority always invests in new facilities if extra users are willing to pay the cost

[1] Transport economists usually talk of the 'generalised cost' of a journey as the weighted sum of time and money costs. The appropriate weight to use for time (the 'value of time') varies with tastes and circumstances. See [13].

[2] A classic exposition of this argument, together with some empirical evidence, is to be found in [14], but for a simpler treatment see [8] chs 7 and 8.

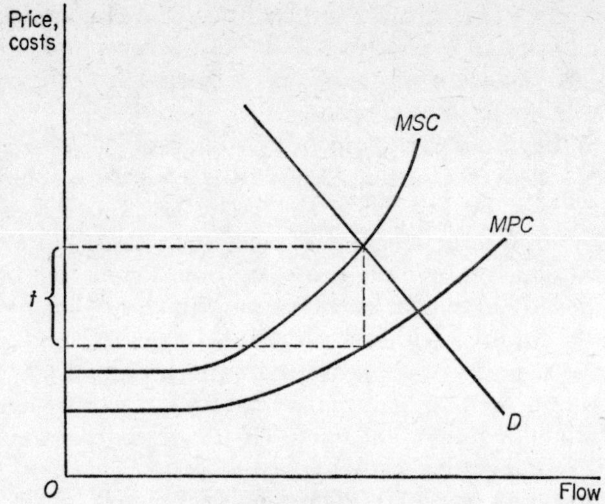

FIG. 1 *Marginal social and marginal private cost*

of these, and that such facilities are perfectly divisible. It will then expand capacity as long as the long-run marginal cost of so doing (comprising annuitised capital cost plus running cost on the marginal journey less running-cost savings on non-marginal units) is less than the short-run marginal cost (*MSC*) of providing one extra journey on the existing road. Thus, in the optimal position, long-run marginal cost will always equal short-run marginal cost (*MSC*). Hence, if long-run marginal cost is constant with respect to flow, and is therefore equal to long-run average cost, then the tax paid by road users as a whole will exactly equal the cost of providing and maintaining the road.

However, that this model represents even a rough approximation to the situation with respect to roads seems highly unlikely [15]. In the first place, road-building decisions are so much the subject of pressure-group politics that it seems unlikely that such decisions would ever conform to this model even if the necessary data to apply it were easily acquired. It is often suggested that economists should advocate the best package of policies known to them regardless of political constraints. But in cases such as this it seems unduly short-sighted to advocate a pricing policy which is wholly dependent for its justification on the adoption of an investment policy that is never likely to exist in practice. Moreover, there may be good distributional or other reasons for departing from the policy suggested above.

Yet even if we accept this assumption regarding investment policy, other assumptions of the model prevent its being applicable to roads. In the first place, there are major indivisibilities and economies of scale in the provision of road space. For instance, it often makes little sense to provide extra width without providing an extra lane, and the cost of building a dual three-lane by-pass is considered by the Department of the Environment to be only 25 per cent above that of a dual two-lane one. Where existing roads are being improved to four lanes, it makes little difference to the cost whether the existing road is two or three lanes wide ([15] p. 185). The construction cost of a four-lane motorway is, on average, some 78 per cent of that of a six-lane motorway [16].

The presence of such indivisibilities does not prevent the application of standard investment-appraisal techniques to determine the 'optimal' level of capacity, but it does mean that capacity will typically be considerably greater or smaller than that implied by the perfect-divisibility model, and the optimal price correspondingly lower or higher, resulting in a deficit or surplus respectively. For instance, consider a model containing a simple capacity constraint, ignoring problems of congestion.

In Figure 2, *MSC* represents the long-run marginal cost of capacity, and Q_1 the level of capacity which would be optimal given perfect divisibility and constant long-run costs. In this case, P_1 is the optimal price. Q_2 represents the actual best position, given indivisibilities, which involves 'excess capacity'. Although the total cost of capacity is greater, in the latter case optimal price, P_2, is lower, that is a deficit is incurred on this particular facility.

Secondly, demand for extra road space can only be satisfied by new capacity in a fairly specific location. In some areas (primarily urban) the costs of providing new road space will be such that congestion will have to be very acute before new road space is provided (and perhaps it will not be even then).

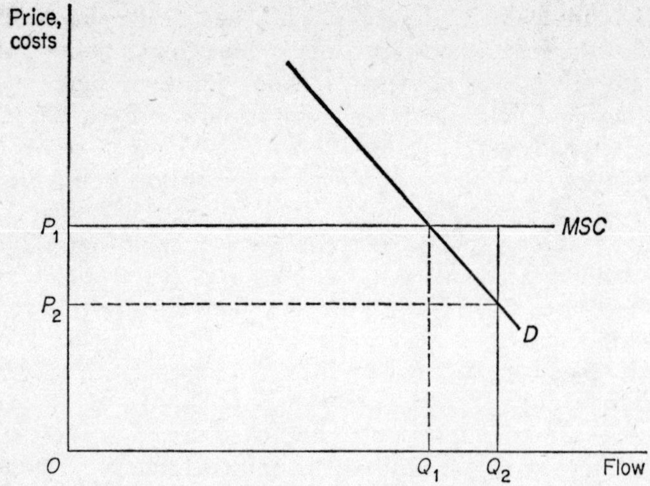

FIG. 2 *Effect of indivisibilities on optimal price*

For such investment is likely to involve demolition of homes, shops and other property. Thus such roads may be earning high surpluses, but still not be worth extending. In rural areas, road space may be far cheaper, and – because of the indivisibility argument – deficits more common. But only wishful thinking could lead to the conclusions that the two effects will exactly offset each other.

So far we have considered the revenue raised from road users as a whole. But the 'track-costs' argument requires revenue to equal cost for each individual group of users. To the extent that there are genuine joint costs, it is not clear which group of users is expected to bear these costs. One argument is that each group should at least bear the incremental cost of providing for it [18]. For instance, heavy vehicles should pay for better surfaces, high vehicles for higher bridges. Yet there is still a major indivisibility involved; for if it is worth building a higher bridge for one vehicle, there are no extra costs involved in permitting other vehicles of equal or lesser height to use it. There may be more flexibility with surface quality – if the flow of heavy vehicles is small, then it may be cheapest to provide a poor surface and bear the extra maintenance costs; as the flow increases, then maintenance costs could become so high that it is worth installing a higher-quality surface to begin with. Even so, it is most unlikely that the marginal cost of extra use by heavy lorries is constant.

Thus we conclude that there is no reason to suppose that 'optimal' road taxes (even in this restricted sense) will exactly cover the costs of providing and maintaining road space. On the contrary; in some circumstances high surpluses may be desirable; in others, losses. It is sometimes suggested that such losses should be recovered by means of a balancing toll, charged proportionately on all transport users. Such a toll may permit the price ratios for alternative forms of transport to equal marginal social cost ratios, although compared with prices elsewhere in the economy there will still be some distortion.

A similar analysis may be applied to the track costs of other facilities, such as railways and canals.[1] To the extent that uncongested facilities already exist, there is no case for eliminating traffic that is covering its marginal maintenance and operating costs. Yet even in the long run, when considering investment in new facilities, there is no reason to suppose that constant returns to scale prevail. Similar problems of indivisibilities and costs varying with location arise. Optimal pricing and investment policies may again imply surpluses or deficits. Thus the dictum that 'every tub should stand on its own bottom' is a poor guide to resource allocation within the transport sector, even before differing external effects on other sectors are taken into account. The cost pattern for transport infrastructure is such that comparisons of costs, revenues and taxes at this level shed no light on the issue of public versus private transport.

[1] Indeed, the European Economic Commission has stated that preparations should be made to charge for these services, as well as roads, on a marginal-cost basis, but with the use of balancing tolls to avoid deficits [19].

3 Environmental Effects of Transport Systems

Whilst the central issue in debates between the proponents of public and private transport in the early years of the rise of road transport was that of track costs, in recent years the emphasis has shifted. The prime driving force behind campaigns for the transfer of traffic from private to public transport, and from road to rail, in recent years has been the desire to improve the quality of the environment.[1] In the case of track costs the difficulty has been to levy appropriate charges for the use of transport facilities; in the case of the environment, the trouble is that, in most instances, no charges at all are levied.[2] Thus the individual or organisation taking the transport decision has no incentive at all, other than social conscience, to take into account the environmental consequences of his decisions (even if he is aware of them). In other words, the environmental consequences of transport systems are externalities arising from the processes of production and consumption, and their presence must immediately arouse suspicion of resource misallocation.

The way in which this misallocation arises will be familiar to all students of welfare economics, and is only summarised very briefly here.[3] In Figure 1 (p. 16) a position was illustrated

[1] For a comprehensive account of the case in non-technical terms, see [20].

[2] It is true that fuel tax acts to encourage conservation of energy and to discourage road transport; but, as discussed in Chapter 4, the externalities generated by motor vehicles vary extensively with time and place, in particular as traffic density varies. Except for the effect of congestion and road lay-out on fuel consumption, the amount paid in tax does not vary in this way.

[3] The factors considered in this paragraph and the next are discussed at greater length with respect to pollution in a companion volume in this series [21].

in which marginal social cost exceeded marginal private cost. The result was that journeys continued to be made up to the point at which the *MPC* schedule cut the demand curve. However, the socially optimal level of demand, if the benefits of a trip are to be judged by the willingness to pay of the user, is given by the intersection of the demand curve with the *MSC* schedule. This level could be achieved by the optimal tax, t. Those trips suppressed by this tax may divert to other modes of transport, or may cease to take place at all.

Yet there is another effect which such a tax may have, where the externalities comprising marginal social cost include pollution as well as congestion. Provided that it is linked to specific outputs of externalities, rather than simply to the number of trips made, the tax will encourage the use of anti-pollution devices as long as they reduce marginal cost – even if they increase marginal private cost (excluding the tax). The feasibility of such a tax is considered in the next chapter, but the point that the optimal solution may involve changing not only the level of output but also the technique of production is worth bearing in mind whether the tax solution is adopted or not.

Finally, it is worth remembering that just as externalities may arise from the processes either of production or consumption, so they may harm either production or consumption. For instance, traffic congestion may increase costs of production where goods have to be transported; traffic noise may reduce the output of office workers. Yet in respect of environmental effects it is usually considered that the consumption disbenefits – often known as amenity effects – are the most significant.

EXTERNALITIES FROM TRANSPORT SYSTEMS IN USE

When one talks of the environmental effects of transport systems, it is to the noise, air pollution and social disruption caused by motor vehicles in use that thoughts usually turn. For it is these effects that are immediately apparent to every member of the community whenever he walks down a street, or – all too often – whilst he is in his home or place of work. The standard approach of cost–benefit analysis to this problem is to attempt to see what sum of money people would on average be willing to pay for a certain reduction in environmental pollution, in order that the extent to which it is worth giving up other material benefits to reduce the effects of traffic may be gauged.

Probably the best researched of these factors is noise nuisance. In the United Kingdom, regulations currently stipulate the maximum noise to be generated by a goods vehicle in use as 92 dB(A), and by a passenger vehicle as 87 db(A), although enforcement is difficult. Noise of this order may create severe annoyance, but what is also important is the duration and frequency of the noise. Thus measurements of noise level are usually stipulated in terms of the level that is exceeded for a given percentage of the time, for example 10 or 50 per cent. Attempts to place money values on noise have usually taken one of two forms. In the first place, the cost of noise-exclusion devices, such as double-glazing, is assessed. This is appropriate as long as inhabitants would prefer to incur this cost rather than suffer the noise in question, although it obviously excludes the disbenefit of noise outside the building; on the other hand, double-glazing may be valued partly for thermal insulation. An interesting development from this procedure has been an attempt to estimate the demand curve for exclusion facilities in the vicinity of Heathrow Airport [22].

The second approach has argued that people reveal their willingness to pay to avoid noise in house-purchase decisions; thus physical measures of noise are related to house prices. However, whilst there is evidence that house prices respond to aircraft noise, no such relationship was found with road-traffic noise in a study by Diffey [23]. This is surprising in view of evidence such as that in the Wilson Report [24], of whose sample of residents 36 per cent stated that they were disturbed by traffic noise whilst in their homes. But the conditions necessary for house-price differentials to reflect accurately people's willingness to pay to avoid noise at home are extremely stringent (see, for instance, the critique in [25] ch. 5), and moreover such an approach excludes the nuisance of noise incurred whilst away from home.

Air pollution from road vehicles takes the form primarily of carbon monoxide, hydrocarbons, oxides of nitrogen, and lead compounds. It is usually considered that these are not normally present in sufficient volumes to be injurious to health [26], although it is suggested that they may be correlated with psychological disorders: in any case, they certainly cause irritation, disfiguring and damaging buildings and plants. As with noise, the benefit of diversion of traffic to public transport would arise from reducing the number of vehicles in use; also, emission of the more dangerous pollutants is lower from diesel than from petrol engines, whilst use of electric traction may eliminate the problem at the point of use. Attempts to place money values on air pollution have largely been confined to costing items such as damage to property, and laundry bills [27], as opposed to valuing the amenity aspect of clean air, although the house-price-differential approach has been used here. Other ways in which vehicles damage property are through vibration and, particularly in narrow streets, accident damage. They also cause considerable visual intrusion [28].

The social disruption caused by motor vehicles consists primarily of delays, diversions, danger and inconvenience to pedestrians, which both boost the cost to them of making a trip on foot and reduce or divert the trips made. Little is known of the significance of such effects, however, as research on transport problems has concentrated, strangely, on the movement of vehicles rather than that of people. Evidence suggests that (not surprisingly) people place a much higher value on time savings gained whilst walking or waiting than whilst in a vehicle [29].

With respect to accidents, the difference in accident rates between road and rail is well-known. For instance, in 1971, there were 31 deaths and 367 serious injuries per thousand million passenger-miles travelled by road in the United Kingdom; the corresponding figures for rail are 8·6 and 12·4 respectively ([20] appendix 3). Whilst this may partly reflect the current allocation of resources for safety measures on road and rail rather than the intrinsic characteristics of each mode, there seems little doubt that the reduction of road traffic by the transference of passengers to public transport and freight to rail would save many deaths and injuries. Some of the costs of accidents – damage to property, loss of output due to injury or death,[1] medical expenses – are in principle readily quantifiable in money terms [30]. But the values placed on pain, grief and death itself are, in current work, largely arbitrary.

If one looks more widely at the effects of private transport, one can include its influence on the location of homes and work places, where the private motor vehicle has encouraged low-density urban sprawl at the expense of the countryside [31]. It should not be assumed that all the external effects of the motor-car are negative, however, This same urban sprawl may create a more pleasant urban environment than would otherwise exist.[2] Car owners often give lifts to others, or bring happiness to friends and relatives by making more visits than they would if reliant on public transport. Such external benefits of mobility should not be overlooked when seeking to reach a balanced view on the transport question. Nevertheless, it does appear that, over all, private transport involves the production of major external disbenefits relative to public transport.

EXTERNALITIES ARISING FROM THE PRODUCTION OF INPUTS FOR THE TRANSPORT INDUSTRY

Most recent writings on transport economics recognise the importance of externalities created by transport facilities in use. What is less common is for transport economists to trace back the production of inputs for the transport industry. Unfortunately, external effects at that stage may also be very significant.

All transport modes need a source of energy, and at the present time the bulk of this energy is supplied by oil. Now the oil industry itself is a major source of external effects, including air and water pollution, visual intrusion, and the destruction of

[1] It is usual here to deduct from the victim's output that part which he would have consumed and consider only losses to his family and society at large. Recently, D.O.E. treatment of this item has been reversed. Yet, surely, the value of the consumption lost by dying has little to do with one's wish not to be killed in a road accident!

[2] Whether this is so is itself hotly debated. The classic attack on the desirability of low-density development with segregated land-use can be seen in [32].

recreational facilities. Even where the immediate source of energy for traction is electricity – usually thought of as a clean and environmentally desirable source of energy – the generation of that electricity and the production of fuel for power-station use, whether the fuel used is oil, coal or nuclear power, involve externalities. The manufacture of transport vehicles also involves the use of inputs of iron and steel, aluminium, rubber, and – both directly and indirectly – electricity, coal and oil. All of these industries are themselves major users of transport inputs and hence contributors to the externalities described above.

Such a pervasiveness of externalities involves great complications for economic analysis. Whilst evaluations of alternative modes of transport are increasingly attempting to place shadow prices on externalities in use, market prices of inputs into the transport industry are usually taken as representing appropriate social valuations of the costs involved. But a comprehensive analysis would require tracing and a valuation of all the externalities involved in the production of inputs as well. The simplest method of trying to disentangle such effects is by the use of input–output analysis, in which current inter-industry flows are used as the basis for estimating the effects on all industries of changes in final demand. A pioneering attempt has been made at this by Victor [33]. Using a sixteen-industry input–output analysis of air, water and solid-waste pollution for the Canadian economy, he found the oil and coal industries to be those inflicting the greatest ecological loss per dollar of output. In an application of his model, which he stresses is exploratory rather than conclusive, he predicted that a switch of 50 per cent of private-car-miles to public transport would result in a fall of 'ecological costs' of 6·5 – 8 per cent of the total for all economic activity. His valuation of ecological costs, however, was not based on the usual approach of evaluating consumers' preferences, in view of the obvious difficulties in this context, but on the relative weights attached to various forms of pollution by a single chemist. None the less, the exact choice of weights does not appear greatly to affect the outcome.

The problems of aggregation and assumed linearity in the use of input–output analysis are well-known, but sadly they appear to be particularly important in the case of the transport industry. Thus the industry is seen in this analysis as a producer of passenger-miles and freight ton-miles, each unit of which requires constant inputs from other industries; but a passenger-mile of transport in an urban rush-hour is a very different commodity from one in a rural area, whilst the inputs needed to produce a freight ton-mile vary not only with location and time but also with the commodity carried. That certain modes are used predominantly for certain types of transport task makes inter-modal comparisons using input–output analysis particularly suspect, for traffic transferred may well be less suitable for the new mode than existing traffic. Nevertheless, progress may be made by looking at specific transport facilities first and forecasting the direct inputs needed for them, and resorting to input–output analysis only to forecast the indirect inputs required. In any event, input–output analysis provides a valuable framework in which to look at the problem in more detail.

It is worth singling out for special attention the provision of transport track, whether road or rail. Inputs required for this again present environmental effects, such as those from the quarrying of aggregate for road construction[1] and the visual intrusion created by the completed track; but a particular problem is presented by the required land inputs. Acquisition of these often involves dispossessing (by means of compulsory purchase) occupiers of residential commercial and industrial premises. Whilst the notion of householders' surplus, that is that amount of compensation over and above the market price that would be necessary for the resident to move willingly, is now familiar [35] (and, indeed, some allowance for this is made in the 1973 Land Compensation Act), measurement is difficult; and the effects on workers and customers of the compulsory purchase of other types of property such as shops, offices and factories, which may deprive them of valuable local facilities or even their livelihood, are largely ignored. Even where land is currently lying derelict, it may well be used for recreation,

[1] Demand for road maintenance and construction accounts for 25 per cent of the total demand for sand and gravel in the United Kingdom, and is growing rapidly. See [34].

and in any case the construction of transport facilities still involves an opportunity cost in terms of the alternative uses to which the land could have been put. In a free market, and in the absence of externalities, the current market value of the land may provide a guide to its opportunity cost.[1] But most uses of land create positive or negative externalities on the neighbourhood, and whilst the necessity to obtain planning permission should limit the cost of the latter, the tendency of transport schemes in built-up areas to take some of the remaining open space undoubtedly deprives inhabitants of beneficial external effects.

DEPLETION OF NATURAL RESOURCES

Alongside concern with the deterioration of the environment in recent years has been a growing anxiety at the rate of depletion of natural resources. However, it is difficult to reach any firm conclusions on the seriousness of this problem, as at any point in time known reserves of a resource may only be a small part of total recoverable reserves, future technological discoveries which will provide substitute materials or reduce costs of recovery are hard to forecast, and there is an important feedback effect from rate of depletion via increased costs of recovery to the rate of growth of demand. Indeed, it is sometimes argued that the price mechanism itself will ensure that depletion takes place at a socially optimal rate, and hence that the market prices of natural resources themselves reflect the full social cost of depletion. In essence, the argument is that owners of natural resources will forecast future prices and will only supply the amount of resources now for which the current price exceeds the present value of the forecast future price [36]. There are many reasons why one may have severe doubts about this as a statement of how natural-resource-depletion rates are in fact determined. Empirical investigation of the investment decision suggests that most firms have very limited time horizons [37], and this may be particularly true of natural-

[1] It should be stressed that the historic cost of purchasing land is no guide at all to current opportunity cost.

resource depletion, where political and technological uncertainties abound. It is not clear that the market rate of discount, by which future prices are discounted before comparing the value of future sales with that of the present, is an appropriate one to use in social decision-taking, given the very low weight that it places on consumption in the distant future. If, in the long run, lack of natural resources does considerably constrain world output, surely such low weights would be inapplicable [38]. In any case, even if the price mechanism did work in the way suggested, there would still be an increase in the price of natural resources over time, as they become more scarce and as less-favourable sources of supply are utilised. Natural-resource prices are also clearly subject to major influence by the international political scene. It is essential that in considering long-term transport investment, such future price rises be taken into account, rather than assuming that current relative prices will still hold in future years.

When discussing resource depletion in the context of transport, it is obviously to fossil fuels, and particularly oil, that thoughts first turn. It has been estimated [39] that road transport alone accounts for 25 per cent of oil consumption (and 10 per cent of all energy consumption) in the United Kingdom. In the United States, the transportation industry as a whole used 55 per cent of petroleum consumed in 1965, and its share was growing [40]. Table 1 illustrates the relative consumption of energy by different modes of passenger transport. It will be seen that the degree of economy in energy use achieved by a switch to public transport depends very much on the load factors attained – the more concentrated the traffic, the greater the saving from switching to public transport. It should be noted that the table refers only to energy consumption for traction purposes, and does not include energy used in producing the inputs for the transport systems.

In the case of freight transport, it is more difficult to generalise because of the variety of commodities and patterns of operation. Given traditional railway operating methods, and taking into account collection and delivery, shunting, and so forth, it may be that in some cases rail transport is *more* energy-intensive than road, although water transport is certainly less [41]. But wherever major flows of traffic are concerned, rail is likely to offer considerable savings.

TABLE 1 *Fuel consumption by transport mode*

	Miles/ gallon	Average occupancy	Passenger-miles/ gallon	Maximum occupancy	Passenger-milcs/ gallon
Motor-car	30	1·3	39	4	120
Rear-engined double-decker bus	7	16	112	75	525
2-car diesel railcar	4	35	140	150	600
3-car electric railcar	1·5	50	75	250	375
London Transport 7-car tube train	0·75	105	79	840	630
Light transit 2-car set	2	50	100	240	480

SOURCE: [39]. The conversion factor of 1kWh=0·67 gallons of oil is used. For details see the original.

Another advantage often cited for public transport over private is the more ready application of electricity as the source of energy. The attraction of this is that coal reserves are believed to be far more plentiful than oil, and moreover most non-fossil-fuel sources of energy, such as nuclear or hydroelectric power, are likely to be used primarily for electricity generation. However, whilst this is likely to be important in the long run, as long as oil is used widely for electricity generation, electrification of a diesel-operated public-transport service will probably boost over-all oil consumption because of energy losses in the course of such a roundabout method of producing traction, although the mix of oil types may change.

Oil is not the only scarce natural resource used in the transport industry. The vehicle-construction industry uses large volumes of resources such as iron, steel and aluminium, for which no early supply problems are anticipated, and smaller amounts of other, scarcer metals, such as lead, zinc and copper. Whilst it is estimated that supplies of the latter three metals – other than from recycling[1] – would be exhausted within the next twenty-

[1] Recycling of these metals from car bodies is already fairly efficient, but obviously cannot permit an expansion of the stock of vehicles using fixed amounts. Lead used as an additive in petrol is, of course, too dispersed to recycle.

five years if present trends continued, it is often suggested that they can largely be replaced in vehicle construction by the use of aluminium and plastics [42]. The result of this may be to increase the rate of energy-resource depletion, since plastics require oil or coal as a raw material, whilst the smelting of aluminium requires vast energy inputs. Again, the necessity to trace the effects of such substitution is demonstrated. Input–output analysis is of some assistance, but in view of the relative price shifts involved in such technological changes, the assumption of fixed input coefficients must be extremely doubtful.

Even if the resource-depletion effects of alternative transport systems can be estimated, it is still difficult to introduce this factor into policy evaluations. Ideally, one would wish for a shadow price for the resources in question, reflecting the opportunity cost of their future use. But long-run forecasting on the scale necessary to estimate such shadow prices is in its infancy, and is a controversial and hazardous business [43]. In their absence, one can certainly say that less resource depletion is desirable, *ceteris paribus*, and one can at least guess which resources are in most need of conservation; but as to the exact trade-off between depletion and transport benefits, little is known.

CONCLUSION

The underlying theme of this chapter has been that, in the transport sector, externalities are so all-pervasive that existing market prices are of little use in guiding the allocation of resources. Unfortunately, many of the effects of transport decisions are both difficult to trace and difficult to value. In part this difficulty arises from the immense variety of tasks that the transport industry performs, which makes generalisation difficult. It also arises from the lack of suitable alternative markets in which to see how people value the type of external effects produced by private transport. In this connection, one encouraging development is the extension of attitude surveys to present their subjects with explicit trade-offs between various types of costs and benefits [44].

Yet when it comes to valuation of effects the problem is greater than simply a fault in the current state of the art. In the case of many important factors mentioned in this chapter, there is reason to suppose that the economist's normal valuation procedure is simply inappropriate. Where general knowledge of the consequences of effects is slight, individual values must be based on faulty information. Where the effects involve suffering – whether physical or mental – few people would think of a straight trade-off with material well-being. In such circumstances economists have turned to sociologists for help, but to expect sociologists to be able to measure suffering in the same units as economists measure material well-being is to ask for the impossible [45]. One just has to accept that there are many important factors which cannot be reduced to a money measure. In this case, to regard the problem as the standard one of externalities is misleading. It is not the absence of markets that is the main problem, nor would the creation of markets in such effects be of much help.

To say this is not to conclude that the role of economic analysis in transport decisions is slight. On the contrary, a rational decision on transport problems will require knowledge of the economic effects of alternative solutions, so that these may be traded, ultimately, with other factors analysed by different disciplines. What is denied is that all the effects of transport systems can sensibly be expressed in a single measure of net social benefit.

4 Pricing versus Controls in the Transport Sector

The previous two chapters have stressed the importance of external effects, and the consequent inappropriateness of market prices based on the money costs of providing different systems in choosing between public and private transport. In such a situation, the solution traditionally put forward by the economist is that the government should intervene with taxes and subsidies designed to bring prices in line with the marginal social costs of using the resources in question. This having been done, the allocation of traffic between different modes would then be left to the individual decision-taker. The government would have no interest in the nature of this final allocation, except for the need to forecast it in determining investment policy.

Now the most obvious problem with this approach is one of measurement of marginal social cost. Methods exist for the valuation of the extended journey time caused by the congestion effect of an additional vehicle in the traffic flow, although these do not allow adequately for the variation in the valuation of travel time with length of trip, amount of time savings, time of day and week, comfort and convenience of the mode in question and variations in individual tastes [29]. With respect to many of the other externalities, valuation methods do not even exist [46].

In view of this, it is often argued that the pricing solution is not feasible, and that planning and administrative controls are more appropriate. This argument appears to confuse the choice of an objective with the choice of a method for its achievement. The previous paragraph suggested that the choice of mode for any given traffic involves wide-ranging effects, some of which may be evaluated using standard cost-benefit techniques and some of which may not. But once a choice has been made, measures to bring the desired traffic allocation about may be either administrative, or may take the form of pricing policy. The choice between the two is largely a question of equity and of administrative costs rather than a matter of welfare economic theory, and it is to these issues that we turn in this chapter. Yet it is also a question of consumer knowledge and rationality; if, because of lack of knowledge, misperceptions or habit, the individual decision-takers do not act in their own best interests in the light of the prices they face, this could be an argument for the imposition of direct controls, rather than leaving individuals free to react to prices. Such an argument will be met again, particularly with respect to freight transport.

EQUITY IN TRANSPORT PLANNING

Most applied welfare economics takes as its criterion of desirability a comparison of the willingness to pay of those affected for the benefits of a plan, and the social costs of the plan measured again in terms of the willingness to pay for the alternatives of which the plan deprives them. However, for a particular benefit willingness to pay clearly depends on income as well as on the individual's tastes. Thus the appropriateness of this criterion rests on two value judgements; that individual preferences are to decide the issue, and that those preferences should be weighted according to the current distribution of market power [6]. It is useless to try to avoid the problem by suggesting that transport decisions are not good ways of influencing income distribution. The fact is that they *do* redistribute income, whether we like it or not.

The equity issue arises to the extent that one considers the current distribution of income an unjust basis on which to distribute transport benefits and costs. The use of transport facilities varies greatly with location and car ownership, and these in turn are closely related to income. For instance, in most U.K. cities there is little doubt that measures to improve the quality of private transport usually benefit primarily the better-off sections of the community, whilst those to improve urban public transport are more likely to benefit the poorer. Exceptions arise with higher-quality or long-distance commuter services, such as rail services into London, users of which are typically better off than other public-transport users. In countries such as the United States and Australia, where car ownership is much higher than in the United Kingdom, much recent public-transport investment has been in rail services of this type, used by relatively affluent central-business-district-orientated commuters. It is more difficult to assess the incidence of improvements to freight transport, since after price adjustments it is not clear whether the manufacturer, the haulier, the retailer or the consumer receives the benefits from lower transport costs.[1]

The importance of this issue for the pricing versus controls debate is that, apart from the possible effects of controls on relative property values, whilst controls on the use of transport facilities affect all potential users equally (unless specific exemptions, such as for the disabled, are made), pricing will usually restrict their use by the poor more than by those with higher incomes. This issue has been raised, for instance, with respect to the desirability of road pricing, although conflicting views have been expressed on the incidence of benefits and costs [8]. The overall conclusion must be that, given the difficulty of generalisation, any specific proposal on transport pricing or investment should be examined with respect to the incidence of its effects as well as to the magnitude of the costs and benefits.

ADMINISTERING OPTIMAL PRICING SYSTEMS

Suppose that there were no administrative costs attached to operating a pricing system. What would a scheme of optimal transport prices look like? Prices would vary according not only to the cost characteristics of the vehicle and route in use,

[1] The issue is not clear-cut even for passenger transport. Better transport services may lead to higher rents and land values; thus property owners rather than transport users may benefit. The identity of these groups may or may not coincide. Similarly, higher wages may partly or totally compensate for increases in transport costs [47].

but also to the time of day and year, since demand – and hence congestion and marginal cost – fluctuate cyclically. For instance, most routes have morning and evening work-journey peaks. Holiday routes incur peaks particularly at summer weekends; routes used by certain types of freight, such as coal or sugar-beet, at other times of the year.

It is fairly commonly agreed that peak fares should be higher than off-peak fares, and attempts have been made to calculate the relative marginal costs of additional peak and off-peak bus services. (See, for example [49]). The standard model [17] for examining this type of situation divides costs into those that are related to volume of capacity (capacity costs) and those that are related to the use to which that capacity is put (running costs). Off-peak price is set equal to the latter cost, whilst capacity is chosen by equating the marginal benefits to peak users with the cost of expanding capacity, and peak price used to ration demand to this level. The solution is illustrated in Figure 3. For ease of exposition, all costs are assumed to be constant per unit of capacity or use. The optimal level of capacity is Q_1 and the peak price which causes this to be fully

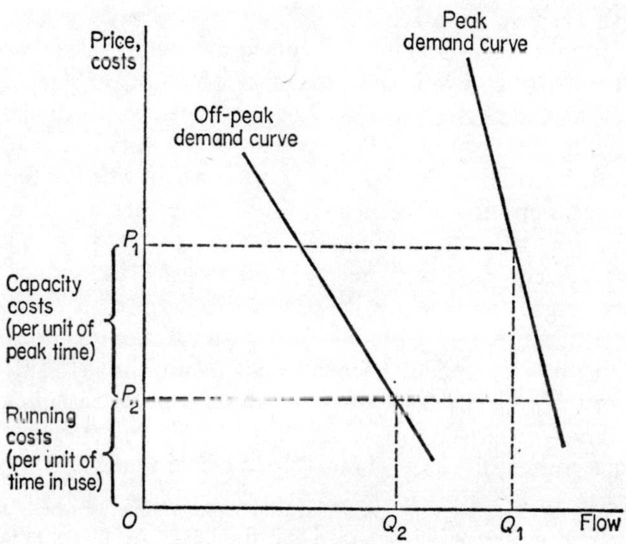

FIG. 3 *Optimal peak and off-peak prices*

utilised is P_1. Any increase in peak output would involve costs equal to P_1 per unit, and users would not be willing to pay this much for the extra output. In the off-peak situation, price is set equal only to the marginal cost of increased use of existing capacity, that is P_2; the output level is Q_2.

In the case of public transport, most of the costs are associated with the provision of a particular level of capacity. Once a decision on this level is taken, the incremental cost of using capacity is very low – often zero – if a service is running with empty seats. Thus one would expect off-peak prices in public transport to be very low, whilst prices for periods in which capacity is in full utilisation would be high. However, in practice, average-cost pricing has been the rule. Political constraints – that this would have the appearance of discrimination against particular types of users – have usually prevented the raising of peak prices, whilst financial constraints – that authorities are unwilling to provide the necessary subsidies – have prevented the lowering of off-peak prices. In any case, the consequences for the distribution of income of such a change in pricing policy would not necessarily be desirable, given that it would require funds to be raised or diverted from elsewhere.

With respect to private transport, the problems are even more intractable. Given that public-transport users are expected to buy a ticket, the administrative cost of varying the fare according to the route and even to the time of day are not necessarily very great.[1] But private-road users are usually charged for road space by a combination of licence fees and fuel taxes, neither of which can discriminate according to time of day, nor, in general, according to route ([15] ch. 7). The pricing system discussed here would require individual tolls, either charged manually, with consequent high costs not merely of providing toll-levying facilities but also in the form of delays to traffic, or charged by means of electronic metering devices, the capital and running costs of which would be substantial [50]. The first system may be worthwhile for specific long-distance routes; the second possibly in certain heavily congested

[1] This is not to say that the administrative costs are always worth paying, especially when considering one-man operation of buses. Flat-sum fares do speed loading in such circumstances.

areas; but it seems unlikely that either would be appropriate in general. It should be noted that complicated price structures are only of value in any case to the extent that consumers can understand them and react to them.

Even worse problems arise with respect to the issue of using pricing to control the emission of pollutants. As mentioned in Chapter 3, ideally this would involve gearing charges for the use of roads to the level of externalities produced rather than to the level of road use. Otherwise, no incentive is given to switch to less-polluting technologies. Even a broad approximation to this would require different taxes for different types of vehicle, according to the forecast mean level of pollution produced. But it would still be desirable to charge tax at the consumption stage, since it is impossible to forecast to what extent, and in what conditions, a particular vehicle will be utilised. The administrative cost of a system such as this may well make the setting of standards a more attractive procedure, although imposing these effectively may not be cheap either.

A number of alternative methods of pricing which permit more limited discrimination by time and route have been suggested, of which the simplest are probably special licences for access to particular areas, and parking fees [51]. Since the latter do not affect through-traffic, they are likely to be most effective when applied to major trip destinations such as central business districts or shopping centres. Obviously, such methods cannot discriminate between routes to the centre, and can only crudely discriminate between times of day; but it should not be thought that this means that the resulting price is purely arbitrary. Charging a common price for products with different costs is common practice in all industries, especially service industries where marginal cost varies with time of day; but all this does is add an additional constraint. If one seeks to achieve the closest possible approximation to the results of marginal-cost pricing subject to this constraint, the solution is to take the weighted average marginal cost, where the weights are the absolute values of the slopes of the respective demand curves ([52] ch. 4). In other words, the more responsive a particular category of demand is to price, the more important it is to get price close to its optimal value for that category.

SECOND-BEST PROBLEMS

So far in this chapter we have given many reasons for supposing that non-optimal prices will be common in the transport sector, and for supposing that once administrative costs are taken into account, such departures from so-called optimal prices are often justified. However, this has important policy consequences.

In the first place, it opens the door to second-best pricing solutions. Now to the extent that prices outside the transport sector are non-optimal, the problem of second-best arises in any case. For in these circumstances it may be that the best policy within the transport sector differs from the policy postulated so far. For instance, suppose that, due to market imperfections, price, on average, exceeded marginal social cost elsewhere in the economy. To the extent that transport and other goods are substitutes, this would justify some excess of price over marginal social cost within the transport sector ([52] ch. 3).

However, divergences between price and marginal social cost within the transport sector have more serious implications for pricing policy elsewhere in the sector, since the strength of the substitution relationship is likely to be much greater. For instance, suppose that a fixed number of journeys to work, Q_3, is made from A to B (Figure 4). The average generalised cost of private transport (AC) rises with volume as congestion increases; the only public-transport alternative assumed to exist has a constant average generalised cost of P_1. For simplicity, assume that this equals marginal social cost. The optimal distribution of traffic with Q_1 persons using private transport and $Q_3 - Q_1$ using public transport, could be achieved by levying a toll (t) on car users. This may be expensive to administer. However, exactly the same modal split may be obtained by reducing the generalised cost of public transport to P_2, either by reducing the fare or by increasing the quality of the service.

Administratively, such an alternative appears very much simpler; however, it does have drawbacks. In the first place, it reduces the generalised cost of transport by either mode, at the margin, from P_1 to P_2. This may increase the number of peak-hour trips from A to B, at a social cost greater than the social benefit from these trips. Secondly, it involves a different distribution of income. Transport users on this route gain, while government revenue is reduced. If the government in consequence raises taxes or prices elsewhere, this may distort resource allocation in other parts of the economy.

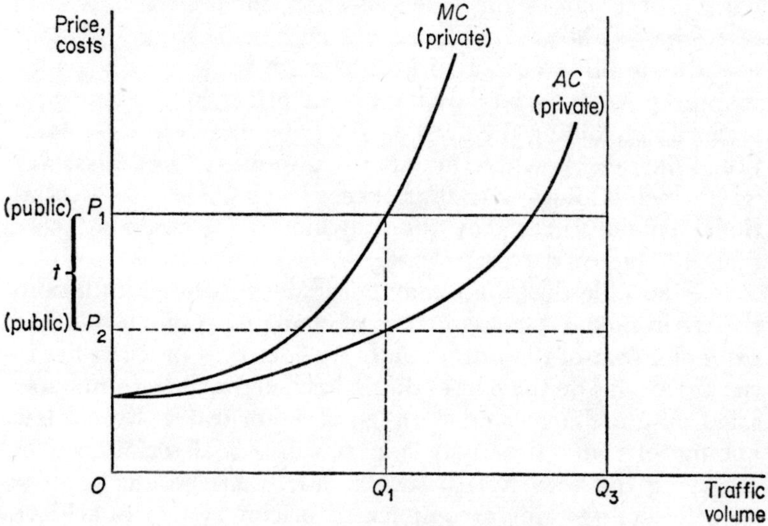

FIG. 4 *Second-best pricing policies and modal split*

Of course, transport services can also be complements. This is the case where one mode acts as a feeder to another. Here the implications are the reverse – if price is low on one mode, this may provide a case for raising it on the other. But it is unlikely that a sufficiently close degree of complementarity for such policies to be worthwhile often exists. For instance, in Melbourne most of the bus services which act as feeders to the state-owned rail system are privately owned. Suppose that they set fares above marginal cost. To the extent that people are discouraged by this from using the combination of bus and rail, one might expect to be able to counter the effect of the bus fare by an equivalent reduction in rail fares. But this may promote rail travel using the private car rather than the bus as feeder, and the social consequences of this may be very undesirable.

In such cases, direct action to reduce the bus fare, by subsidy if necessary, is likely to be better.

The problem of second-best has relevance in the argument between the use of prices and controls as well. Suppose that by means of controls, for instance closing certain areas or streets to private traffic or by giving priority to public transport in the pattern of junctions and lane allocation, the desired allocation of traffic as shown in Figure 4 could be achieved without lowering public-transport prices. The advantage of this policy in preventing a general reduction in prices in the transport sector and avoiding the need for the government to raise additional finance elsewhere may outweigh any reduced accuracy with which it allocates traffic between modes. Also, its distributional consequences may (or may not) be considered more desirable in any particular case.

It is possible that there may be significant indivisibilities in the environmental consequences of changing traffic levels. For example, part of the attraction of a pedestrian precinct lies in the knowledge on the part of the pedestrian that he is completely safe from the danger and intrusion of motor traffic. Even a low volume of traffic flow may impose substantial social costs in removing this security. In such a case, there would be little point in constructing a complicated pricing system to achieve a traffic flow of zero; an outright ban will do the job at far lower cost.

CONCLUSION

The outcome of this chapter has not been to suggest that any single policy of adopting new methods of charging for road space, of reducing public-transport fares, or of restrictions on the use of private transport, will be appropriate in all circumstances. What has been shown is that each of these policies, or any combination of them, may be the best way of dealing with any particular situation. Transport plans for particular areas and types of traffic need to be drawn up taking into account the wide variety of options available, and assessing the costs and benefits of each. Pricing policies themselves need to be subjected to a wide-ranging cost–benefit analysis, looking not only at their effects on resource allocation, but also at effects on income distribution, feasibility and cost of implementation. For this reason, it is impossible to provide a blueprint for the reform of the transport sector. What we have sought to do so far in this book is to bring to mind the relevent considerations in choosing between alternative divisions of traffic between public and private transport. In the remaining chapters, we attempt, using the evidence currently at hand, to look briefly at the arguments as they apply to specific areas of the transport sector.

5 Urban Passenger Transport

Without doubt, the area of the transport sector that has produced the greatest research effort and the largest number of long-term plans in recent years has been that of urban passenger transport. Many of the problems of congestion and environmental degradation are already apparent in most urban areas; yet land-use/transportation studies typically point to a need to cater for two or three times the current volume of private-car traffic within the next fifteen or twenty years. One major study saw a solution to the dilemma in terms of providing a new hierarchy of urban roads – primary, district and local distributors – in between which would exist 'environmental areas' in which the level of traffic would be strictly limited and confined to vehicles requiring access to property [53]; yet this solution has been estimated by Foster to cost some £18,000 m. at 1964 prices (in [54]), as well as involving massive compulsory purchase and demolition of property.

Unfortunately, the consequences of *not* providing for such a growth in private urban transport are not well understood. All demand forecasts are conditional upon the quality and cost of car- and public-transport services available. If new road space is not made available, increasing congestion will cause the cost of private-car journeys to rise rapidly, leading people to divert to other modes of travel (particularly if a segregated public-transport system is available), and to modify their patterns of trip-making, perhaps changing the location of work places, residences and shopping areas. In other words, congestion itself will constrain the rate of traffic growth, and the complete breakdown of road transport that is sometimes imagined, with speed falling to zero, will never occur (except as a result of an unpredicted dislocation, such as an accident); for no one would attempt to make a private-car trip in such circumstances. For instance, private-car traffic in Central London has grown very slowly in recent years [51, 55]. All that this illustrates is that growth of traffic on the forecast scale is not inevitable, and is not intended to suggest that such a 'do-nothing' policy is desirable. But it is an option that has to be compared with the provision of extra road space, the diversion of traffic to another mode, or the restraint of traffic by road pricing or other means. In some cases, all of these alternatives may be more costly than tolerating a certain degree of congestion (particularly if this is 'controlled', in the form of queues waiting at traffic lights to get on to the main roads, rather than queues on the main roads themselves).

CHOICE OF MODE FOR THE JOURNEY TO WORK

In these circumstances it is not surprising that considerable efforts have been made to discover the major factors that influence mode choice and their relative significance.[1] Since it is in the morning and evening week-day peak, brought about by the concentration of journeys to work, that the most severe congestion occurs, most of these studies have concentrated on the journey to work.

One popular explanation of the rise of private transport for work journeys lies in the so-called spiral of public-transport fares. Public transport is a relatively labour-intensive industry. Crew wages alone typically account for nearly 50 per cent o bus-company costs in the United Kingdom and more in countries with higher real-wage levels such as the United States and Australia. In the case of rail, lower crew costs may be offset by the need for a larger administrative and maintenance staff [57]. This, by itself, inevitably means that in a period of rising real wages, the relative price of public transport to other goods will increase.

In the case of private transport, by contrast, the direct labour is provided by the user himself, and the major inputs – vehicles and fuel – are produced by capital-intensive plants, with the result that, until recently, their real prices had tended to fall through time. Obviously, the recent major increases in the price of fuel – a much more significant component of private

[1] For a recent review of the state of the art, see [56].

than of public-transport costs – have served to act in the opposite direction.

The element of a spiral arises because of the economies of scale present in the provision of public transport. Suppose that fares are raised to recover the extra costs from rising real wages. If this decreases demand, in the absence of a reduction in service frequency there will typically be no cost saving to offset the fall in revenue. Thus a further fare increase is necessary if the company is to break even. A similar spiral may be set off by a shift in favour of car ownership at existing prices. The reduction in public-transport patronage again leads to either a reduced service or a fare increase, which in turn loses more custom.

No doubt, this cost trend is a factor in explaining the change in choice of mode; but studies suggest that it is by no means the most important one. A number of studies of mode choice have suggested that commuters are willing to pay the order of 25p per hour to save time.[1] In this case, a difference of 5p in the money cost of public and private transport will be regarded as equally important as a difference in journey time of twelve minutes. Moreover, such studies show that travellers place much more weight on time spent walking or waiting than on time spent travelling in a vehicle [58]. This counts against public transport. Finally, private transport offers greater comfort and convenience. Many factors – cleanliness, freedom from constraints of time-tables, and so on – contribute to this, but the most important single item other than time and cost is probably the certainty of getting a seat.[2]

In the case of rail, the likely advantage over the car in terms of in-vehicle journey time will often be more than offset by the journey time to and from the stations, unless residences and work places are of relatively high density or the journey is long. Bus yields reduced access time to the main mode of travel, but at a cost in terms of in-vehicle journey time. As long as the speed of buses is geared to the general speed of the traffic flow,

[1] This is the figure given for adult non-working time in [58], revised to 1973 prices but not allowing for the suggested rise of 3 per cent per annum in real terms.

[2] For an investigation of what constitutes comfort and convenience, see [59].

it will be affected by the build-up of congestion just as will the private car. Indeed, the existence of walking and waiting time in bus journeys will mean that in such cases the private car will always remain quicker, unless the user has to park at a greater distance from his destination than the bus stop. In the past, the requirement that new places of employment should provide a minimum number of car-parking places has reduced the likelihood of this being the case.

The imposition of this externality on bus users by car users has serious consequences. Within the United Kingdom, some 45 per cent of motorised journeys in the larger urban areas are still made by bus [60]. Such users tend to come from poorer sectors of the community than do car users; they suffer not only slower journeys but also higher fares (since congestion increases the amount of capital and labour needed to provide a given frequency of service) and less-reliable services as a result of congestion. Since a bus is believed to have approximately the effect on traffic flow of three cars, whilst carrying far more than three times as many passengers,[1] the amount of extra road space or the degree of restraint of other travellers that would be needed to provide congestion-free journeys for bus travellers is typically much less than for the private car.

Yet one can go further. If a proportion of those travellers currently using cars shift to buses, the resulting reduction in traffic flow (measured in terms of passenger-car units) may be sufficient to lead to faster journeys even for those who diverted from car to bus. A study on these lines for London suggested that benefits of up to £20 m. per annum in reduced vehicle operating costs and users' time (valued by the standard procedures) could be achieved by diverting car commuters to buses [61]. In this case, former bus travellers, and car travellers with over ten minutes walking time from their parking space to their ultimate destination, benefited; other former car commuters were made worse off. Also, those diverted may incur unquantified losses of comfort.

However, such a diversion of traffic is unlikely ever to come about by individual action. For each individual, the generalised

[1] Average occupancies for London peak-hours, quoted in [51], are 1·4 for cars and 27 for buses.

cost (that is money plus time) of using a car may be lower than that of using a bus. The situation is illustrated in Figure 5, in which the over-all number of trips is assumed to be constant. As the proportion of traffic using car transport increases, the build-up of congestion leads to increased marginal social cost of journeys by both car and bus. But in Figure 5 the average (generalised) private cost for the car is always beneath that for the bus. Thus there is always an incentive for the individual to switch to the car. In the absence of captive public-transport users, the car share of mode split would rise to 100 per cent. Yet all would receive journeys at lower generalised cost if they all travelled by bus. In the case illustrated, assuming that all relevant considerations have been quantified in the marginal social cost curves, the optimal mode split is at Q, where the external costs of a passenger diverting from bus to car exactly outweigh the difference in private cost.

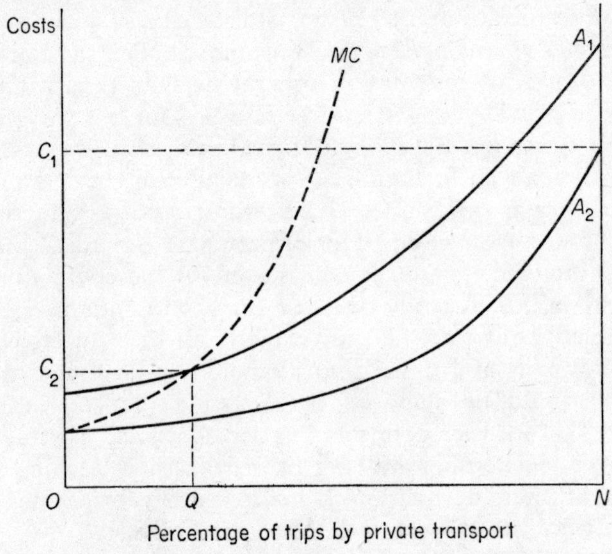

A_1 = average private cost by bus
A_2 = average private cost by car
MC = marginal cost to all passengers of an extra passenger diverting from bus to car

FIG. 5 *The effect of congestion on transport costs by car and bus*

The reader should be warned that this is a gross simplification. In any particular example, some car users may lose by being diverted to bus because they live away from the bus route, or because they place a high value on the comfort and convenience of car travel. As usual, there is no substitute for a case-by-case study of practical situations to discover where users might benefit, on average, from a change in modal split.

Other methods of urban transit – underground and surface railways and other rapid-transit systems – generally have segregated track, and thus do not suffer from the effects of traffic congestion except on feeder roads. Even so, diverting travellers to these modes will affect the level of congestion suffered by remaining road users. For instance, considerable benefits (some 35 per cent of all quantified benefits of the scheme) were forecast to accrue to remaining road users in London from the construction of the Victoria Line underground railway [62].

None of the foregoing argument proves that, even within a given ethical framework for measuring social costs and benefits, diverting traffic from private to public transport will be a better way of relieving peak-hour urban congestion than constructing new roads. It does suggest that a failure to examine such alternatives will lead to gross overstatement of the benefits of such new construction. In the few cases where careful comparisons of the costs and benefits of such alternatives have taken place, such as the studies of Edinburgh [63] and Stevenage [64], public-transport (primarily bus) based strategies have tended to come out favourably.

This is not surprising in the case of typical U.K. urban areas. Where an area is already heavily congested, provision of extra road space may generate a large amount of extra traffic. Although these new users gain some benefits, the benefits to existing travellers from reduced congestion and to residents seeking an improved environment alike are greatly reduced.[1] Fitting new roads into the urban environment typically involves large-scale property acquisition, which is not only expensive to the road-building authority, but also involves major social costs

[1] For a case study of a proposed new urban motorway which appeared of doubtful value once all the indirect effects were taken into account see [65].

of forcing inhabitants to move their homes, shops or work places [66] or deprives them of scarce open space.[1] Segregating the roads may involve redevelopment of a large surrounding area; where – as is typically the case – this is not done, the new road may itself impose severance, noise and pollution on a formerly unaffected area. New roads are perhaps most likely to be justified where they remove through traffic altogether from an urban area. For then they can often be situated outside the urban area itself, and other measures can be taken to prevent the filling-up of the released urban road space with generated traffic. But even then, environmental costs can be high (particularly if recreational green-belt land is affected) and each case needs considering on its own merits.

The issue is less clear-cut in the case of less-dense urban areas, such as are found in the United States and Australia, particularly where work places are less concentrated in specific locations but are becoming increasingly scattered around the periphery of the built-up area. In such cases, where traffic is less concentrated on specific routes, the cost of providing public-transport services of a given quality will be higher, whilst the social costs of car travel are lower. The arguments of this section are strongest in these cities when applied to the journey to work in the central business district or inner suburbs.

The implication of Chapter 3 was that many of the benefits from a switch from private to public transport would accrue not to transport users at all but to other persons affected by the environmental costs of road construction. Again, there are other ways of reducing these costs. One possible method, of segregating traffic from people, for instance by bypassing built-up areas, has already been mentioned.[2] Other ways of reducing

[1] The high cost of construction (some £10 m. per mile at 1970 prices) is what made the projected rate of return on the planned inner London motorways appear low [67].

[2] In many areas the major-roads programme may have such an effect. This is less pronounced in the case of motorways, which because of their limited access cannot cater for local traffic, and in some cases – such as the M3 near Winchester – may attract extra traffic through the town to reach the interchange. In any case, the provision of major new roads rather than more modest bypasses depends on the allocation of inter-urban traffic, which is considered in the next two chapters.

the environmental costs of private transport may be broadly divided into changes to buildings, such as double-glazing, and changes to vehicles, better silencing, the use of air-pollution-control devices such as thermal or catalytic exhaust reactor systems.[1] Obviously, these can only offer partial relief. Perhaps the most effective change in vehicle design from the point of view of improving the urban environment would be the substitution of electric for diesel or petrol traction. In the short run, the most practical way of doing this appears to be still to divert traffic to public transport; in addition to electric trains and rapid-transit systems, battery-electric buses are now coming into use,[2] and it is hoped to improve range and reduce operating costs by substituting new sodium and sulphur batteries for the heavy traditional lead-acid ones. In the long run, the small electric car would appear to be a practical proposition, particularly for urban travel. Unfortunately, electric traction may itself actually worsen pedestrian–vehicle conflict by its very quietness and manoeuvrability.

METHODS OF ALTERING MODAL SPLIT

Suppose that, after these considerations, the case for diverting some urban-work trips to public transport is accepted. Two methods of changing modal split have already been suggested in Chapter 4. The first is a full system of road pricing. Even if bus users were charged for the congestion and pollution buses cause, as they should be in a full-pricing solution, the increase in fare per passenger would be a great deal less than for the private car. Failing this, it was suggested that, as a second-best solution, public-transport fares could be reduced. It should be remembered that in the off-peak period public-transport fares are – because of financial constraints – typically inefficiently high. The evidence for this is idle capacity and the low occupancy of seats on off-peak buses and trains. The additional cost

[1] For a discussion of the economics of alternative ways of reducing traffic noise, see [68]. On the question of air pollution control see [69].

[2] In the United Kingdom the first major application is planned for Greater Manchester.

involved in transporting a greater number of passengers would be negligible. However, it is with the peak that we are most concerned in this section. In this period, it may be that, ignoring the externalities involved, public-transport fares are inefficiently low. For catering for extra journeys here implies either an increase in what is often already gross overcrowding, or the purchase of additional capital equipment and hiring of additional staff. However, it is still possible that a lowering of peak fares may be desirable when the external effects on road users and the environment of any alteration in modal split are taken into account.

What would be the effect of such a reduction in fares? A recent survey of U.K., West German and U.S. experience [70] suggests that, on average, the price elasticity of demand for public transport is as low as about 0·3, and that the cross price elasticity between car and public transport is even lower. It was estimated that even the complete abolition of public-transport fares would lead to a diversion of only some 15 to 20 per cent of car users to public transport.[1] Other extra public-transport trips would be made by former walkers or persons not travelling at all.

The total financial cost of adopting a free-fares policy is high – one estimate suggests some $75 m. per annum in the case of Boston, Massachusetts and $2 billion for the United States as a whole at 1967 prices [72]. However, a major part of this sum comprises transfer of income, and it is misleading to regard this as identical to the use of real resources. For instance, for Boston, only $8·6 m. of the total was expenditure on extra capacity to cater for the increase in traffic, whilst this was partly offset by a saving of $3·5 m. from eliminating fare collection. The size of this latter item indicates another major indivisibility; whenever attainment of optimal modal split suggests low fares, it will be worth abolishing fares altogether unless the cost of providing extra capacity for the traffic so generated exceeds this saving. It should be noted that abolishing fare collection may yield other benefits in the form of reduced queueing time; in the case

[1] One early study implying even less diversion of trips [71] used assumed values of time based on the wage rate rather than on empirical investigation, and these were probably far too high.

of one-man operated buses, these benefits extend to passengers already on the bus, and – where the bus blocks part of the road space – to those in other vehicles as well.[1]

It has been suggested that improvements in service quality are a better way of influencing modal split than a reduced – or free – fares policy [70, 72]. In the case of Boston [72], this is the result of comparing the two policies in terms of the necessary financial subsidy rather than in terms of the extra real resources needed. If the income transfer is desirable, or at least not harmful, in itself, and if the costs of raising the additional taxation are low, then the free-fares policy turns out to be the best. It is likely to be political and financial constraints, and the consequent high value placed on public revenue, that lead to the rejection of the free-fares approach.

If it is decided to improve urban public transport, what mode should be adopted? The initial cost of constructing new suburban railways is high. However, many cities already have extensive rail networks. In this case, the marginal cost of catering for additional traffic – except where lines are already working at capacity, as in South London – is usually low. Improved inter-modal connections, new feeder services and the construction of 'park and ride' stations may then be worthwhile overall, even if individually they seem unprofitable. Even so, the possibility exists that, in some cases, a bigger contribution to the transport system of the city could be made by using the lines for segregated tram or busways where the same vehicle can act as its own feeder on ordinary streets.[2] One such conversion in the

[1] Average stop times for two-man operated buses on urban routes with an average of three people boarding and three alighting at each stop have been estimated to be eight seconds; for one-man operation, comparable figures range from eleven to twenty seconds, depending on the type of vehicle and on the fare-collection equipment installed [73]. Some of the benefits of the free-fare system may be achieved by simplifying fare structures; for instance, adopting a flat rate regardless of distance. Unless set very low, these will lose much short-distance traffic as passengers will prefer to walk. This, however, may be a desirable result in the peak, given the high cost of catering for marginal passengers.

[2] The case for this is argued in [57], and although Smith has been accused of taking an unduly optimistic view of the relative merits of bus for London [74], the case may be stronger when applied to smaller cities.

United Kingdom is the planned use of the Newcastle suburban rail system to run a light transit system more resembling trams than conventional trains. In other cases, Runcorn for example, completely new segregated busways are being used.

In many cases the denser pattern of services and higher frequency that can be offered for a given outlay by conventional bus services would make them a more attractive proposition than new railways if their disadvantages in line-haul speed, reliability and comfort could be overcome. To a considerable extent, the first two can be achieved by freeing the bus from the effects of congestion (by systems of bus priority at junctions, reserved bus lanes and/or restraint of private-car traffic) and by modifying patterns of service with express- and limited-stop services as well as the conventional stage-carriage service. Developments on these lines are now fairly widespread, whilst comfort is also being increased in new-vehicle designs. More ambitious are new types of service, such as dial-a-bus in which passengers can be collected from their door by telephoning for the bus to call,[1] or personal rapid transit, in which small computer-controlled rapid-transit cars follow a route selected by the passengers. But it must be said that the high capital and operating costs per passenger of such systems make them very doubtful candidates for many practical applications.

An alternative to making public transport more attractive as a way of altering modal choice is to make private transport less attractive. This can be done by reducing the road space available for the private car, by limiting access to the city centre for private cars by a system of queueing at traffic lights, as is proposed for Nottingham [75], or by reducing the number of parking spaces available (usually associated with the rationing of these by increased charges). Which method is preferred can only be determined by examining effects on transport costs, the urban environment, and administrative costs for specific areas. But the over-all advantage of this type of policy, as opposed to public-transport improvements, is that it forces up the generalised cost of urban transport instead of reducing it.

[1] Experiments with such services are proceeding in a number of U.K. towns, such as Maidstone, Abingdon and Harrogate, as well as in some areas of London.

Low transport user costs that do not reflect the externalities involved have been a factor in promoting commuting to work, both by rail into cities such as London, and from low-density suburbs by car into other urban areas. Raising the price of urban transport may play its part in discouraging long journeys to work. Moreover, persuading existing car commuters (from the outer suburbs to city centres) to divert to public transport by making that an even more attractive alternative than the private car will give extra benefits to an already relatively well-off sector of the community.

Finally, in the planning of long-run urban expansion and redevelopment, the influence of land-use patterns on modal split must be borne in mind. This does not mean that cities should be designed solely in such a way as to encourage use of public transport. Obviously, the quality of the urban environment depends on other factors as well. Indeed, planning to reduce the mean length of journeys to work may reduce the harm done to the city by the motor-car without bringing about any increase in public-transport trips.

NON-WORK URBAN PASSENGER JOURNEYS

So far in this chapter we have concentrated entirely on the journey to work. There is good reason for this, in that it is here that the case for a change in modal split is strongest, and that the means to achieve this are most readily available. This is not to say that it is only in the peak that congestion occurs. Much of the benefit forecast to accrue to road users in London from the construction of the Ringway motorways was for off-peak non-work journeys. Whilst many non-work journeys are made to and from city centres for shopping, personal business and entertainment, generally one expects to find a more diverse pattern of trips, including visits to local shopping centres, parks and other recreational facilities, and to the homes of friends. Such trips are not well-served by radial public-transport routes.

One question worth asking is whether it is desirable that people make many of these trips at all. It is clear that such trips are heavily influenced by the transport facilities available; for instance, car owners make far more non-walking trips than non-car owners.[1] The usual argument is that they clearly regard the benefits from such journeys as being worth the costs, or they would not travel. There are two main problems with this. The first is again that of the divergence between social and private cost. However, if the roads are uncongested, the divergence is greatly reduced – indeed it may be more than covered by fuel tax. The second is that, apparently, people typically under-perceive the marginal cost of covering additional car mileage,[2] and look only at the cost of fuel. However, once the single indivisible decision is taken to own a car, evidence suggests that the additional capital and maintenance cost involved in running additional mileage is small [77]. The nature of the car-ownership decision, in which a commitment to a high level of overhead cost is made regardless of the mileage run, causes considerable difficulties in analysing urban transport. The decision is likely to be taken on the basis of the relative merits of public and private transport for all journeys, peak and off-peak. However, once a car is available, it is then only the marginal private cost of using it for peak journeys that concerns the user in choosing his mode. On the other hand, if the husband uses the car for work, it probably ceases to be available to his wife (even if she can drive) for her own work or day-time non-work trips. This may influence his choice of mode for the journey to work, or, if the household is sufficiently affluent, lead to the purchase of a second car. Similar high overhead costs exist in public transport, so that the marginal cost of carrying extra off-peak passengers is often zero; but this is not usually reflected in fares.

The consequence of all this is that the peak and off-peak cannot be regarded as entirely separate markets. Improvements in off-peak public transport may either increase or reduce peak car use and vice versa.

[1] See [76]. One has to beware of assuming a single direction of causation; it is quite plausible that the number of trips made influences the decision to own a car, as well as vice versa.

[2] Obviously, if true, then this would also bias their choice between public and private transport for work journeys.

Many would consider the most important problem of off-peak urban transport to be the plight of the large proportion of the population who, for one reason or another, do not have a car available to them [78] and who often face poor or non-existent public transport. Besides non-car-owning households (still nearly 50 per cent of U.K. households), this includes children, the elderly and wives whose husbands use the car for work. However, in the United Kingdom at least, this is a problem more acute in small towns and villages than in major urban areas, and we postpone consideration of it until Chapter 6.

CONCLUSION

The outcome of this chapter has been to show that the choice in urban passenger transport is not solely to build more roads or to suffer the effects of increasing congestion and a deteriorating environment. Many other policies are available; these include reducing public-transport fares and improving services, restricting private transport and increasing charges, and modifying land-use planning to reduce the demand for private transport. The best combination of these policies for a particular urban area can only be determined in the light of the particular circumstances of that town or city.

However, in view of the relatively low cost of public-transport improvements in comparison with the construction of new urban roads, the large number of existing users who would benefit and the small but significant diversion of passengers from the private car (if extra charges or restrictions were placed on private-car use this would, of course, increase), it is surprising that improving private rather than public transport played such a dominant role in most urban transport plans in the United Kingdom throughout the 1950s and 1960s. A number of explanations may be put forward for this. Public transport has often been outside the control of local authorities,[1] and subject to undesirable financial constraints.[2] Bigger central-

[1] Towns with their own transport services appear to have received better services, although often at the cost of large deficits, than those relying on separate companies. See [60].

[2] As an example of the undesirability of imposing profitability constraints

government grants have generally been available for road-building than for public-transport improvements [80]. Transport planning has consisted largly of forecasting by projecting rising trends in private transport into the future and then making provision for them, instead of formulating and evaluating alternative transport plans. However, a change in emphasis is now taking place. The requirement on local authorities to produce structure plans for their own area is leading to an increased examination of the broad options in terms of land-use and transport facilities available. The system of giving grants for transport improvements has been modified to give a block grant for an over-all package of 'transport policies and programmes' submitted to the Department of the Environment.[1] New urban railways are under construction in London and Liverpool, plans are at hand to improve existing services in cities such as London, Birmingham and Manchester, and many cities are studying ways of improving bus services. There is a long-run danger that, as with road schemes, public-transport schemes which are attractive on engineering or other grounds will go ahead without a proper evaluation of the costs and benefits of the alternatives. In particular, too many resources may be devoted to the construction of new rapid-transit systems as opposed to improving – or indeed preventing a decline in – bus services. But it is clear that the case for providing for much urban passenger traffic, particularly work journeys, by public transport is now widely recognised. For instance, a recent report of the Expenditure Committee of the House of Commons concludes that 'There should be a major and substantial effort to improve public transport. Use of private cars for the journey to work should be severely discouraged in areas where it impedes public transport' ([80] p. 19).

---

on public transport, one can hardly do better than return to the example of the Victoria Line, where it was estimated that 50 per cent of the benefits of construction were lost by subsequent increases in fares on underground lines as a whole [79]. The increases in question were to cover the extra money costs imposed by the Victoria Line.

[1] Although, at the time of writing, attempts to hold down government spending may mean that resources for public transport improvements may be reduced as well as those for urban road-building.

6 Inter-Urban and Rural Passenger Transport

Although the external effects of transport systems have been most discussed in the context of urban transport, when we turn to the inter-urban and rural sectors of the market we find a similar range of effects; it is their frequency and magnitude that may differ. In particular, the number of residents whose environment is affected per mile travelled is likely to be, on average, much smaller than for urban trips, although it should be mentioned that most of the trips dealt with in this chapter will have their origins and/or destinations in urban areas. Thus, whatever measures are taken to influence urban modal choice will have some effect in this sector as well. But even outside towns transport systems still deplete natural resources, still cause accessibility problems (by severing footpaths and dividing up farms for instance) and still create noise, air pollution and visual intrusion. People affected by the latter will include many who have deliberately travelled from urban areas in search of peace and quiet, and the conflict is likely to be worst in areas of natural beauty such as the coast or national parks.

PRICING AND INVESTMENT ON INTER-URBAN ROUTES

Suppose that a full system of marginal social cost pricing were implemented on inter-urban roads. The prices charged would probably be even more diverse than on urban roads. Some new motorways with spare capacity, careful alignment and segregation from residential areas could be used relatively cheaply. For it should be remembered that it is only marginal externalities that are relevant here; effects of the motorway structure, as opposed to increases in traffic flow, are not relevant if the motorway would exist in any case and if spare capacity exists. On the other hand, many inter-urban roads are heavily congested either for most week-day day-light hours or at specific time (summer Saturdays on holiday routes, for example). Often these older roads pass through the centres of villages and towns. On these, charges would be high.[1] Of course, new investment may be justified to replace such roads, although here all environmental effects of the new road, not just the marginal effects of increases in traffic, are relevant costs to be set against the benefits achieved.

Thus it is not clear whether, even in the inter-urban sector, motor taxes on average fully charge for the externalities created by additional road use. Now consider a given inter-urban rail network: it is more than likely that, on average, inter-urban rail fares exceed the marginal social cost of rail-use, at any rate outside the morning and evening peaks, and on some routes the peak at summer weekends. Whenever a train runs regularly with a significant number of empty seats, benefits that could be given at near-zero marginal cost are being wasted. In addition, investment appraisal based on comparison of all costs and benefits might indicate the desirability of increasing the number of services run, whereas commercial criteria rule this out because of the reduction in fares to existing users which is necessary to attract more passengers.

An attempt to fill spare seats on a revenue-maximisation basis is often made by offering special tickets, such as day or weekend returns. These are only valid at times which are inconvenient for business travellers, who are believed to have inelastic demands and can therefore be charged highly. It should also be noted that major reductions in inter-city rail fares, given

[1] The possible inefficiency of introducing toll-roads is well illustrated by this example. Tolls are most readily collected on motorways; yet they may then have the effect of diverting traffic on to older main roads where the externalities created are much higher. Even where motorways are becoming congested, tolls levied on them with no corresponding charge on alternative main roads may make matters worse. This is a typical example of the problem of second-best in the transport sector.

that in the United Kingdom such passengers typically come from better-off sections of the community, may well have anti-progressive distributional effects. In any case it appears that for most inter-city travellers at present, differences in journey time are more important than price in determining choice of mode. For instance, it has been estimated that the elasticity of demand for rail travel with respect to time on routes from London to the Midlands and the North-west is as high as 1·47 [81], although many of these trips are newly generated rather than diverted from other modes. The hypothesis that journey time is of greatest importance is consistent with the finding that rail dominates the middle-distance routes of 150–200 miles, whilst the car is more important on shorter journeys and air on longer journeys (Table 2).

TABLE 2 *Modal split on certain inter-city routes*

	Rail	Coach/ Bus	Air	Car
	(in percentage terms)			
March 1966 (before railway electrification)				
London–North-west	59·3	3·9	22·5	14·3
–North Midlands	53·3	2·7	0	44·0
–West Midlands	56·2	4·6	1·3	37·9
October 1966 (after railway electrification of North-west route)				
London–North-west	70·0	4·8	11·2	14·1
–North Midlands	58·2	1·4	0	40·5
–West Midlands	50·9	4·8	2·1	42·2
1969				
Edinburgh–Glasgow	49·0	6·0	0	46·0
Manchester–Liverpool	46·0	3·0	0	51·0
1972				
Ipswich–Norwich	32·9	2·1	0	65·0

SOURCES: [81] for 1966 figures, [82] for later figures.

A major consideration here is the length of the feeder journey plus any waiting time at the terminal. On most main routes air is faster than rail, and rail is faster than the car for the trunk part of the journey, but the car avoids the need for a feeder journey, whilst on average feeder journeys are longer for air than rail. The outcome in terms of over-all journey time is illustrated in Figure 6.

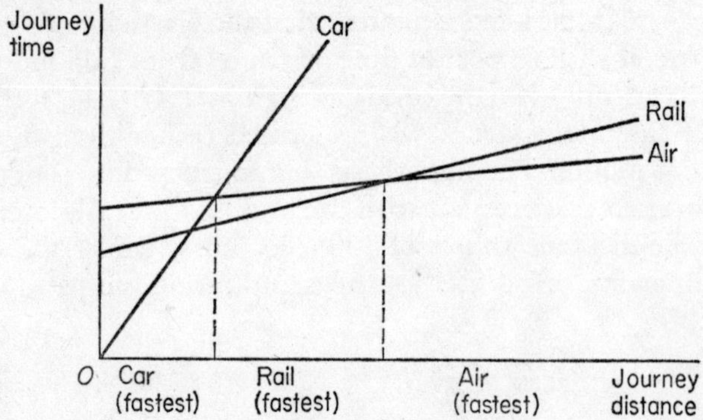

FIG. 6 *The relationship between journey time and distance for alternative modes of inter-urban travel*

The implication is that decisions to invest in improved quality of public-transport service by accelerating trains, increasing service frequencies or improving feeder services will be significant determinants of modal split. However, whilst in the case of road construction such decisions are, supposedly, the result of a comparison of all user benefits with capital cost [83], in the case of rail inter-city routes they are, supposedly, based on estimates of the extra revenue generated. Gains in consumer surplus are ignored. Thus a project is more easily justified under the former than the latter criterion [84]. As an example, it has been suggested that if time savings resulting from remodelling the rail layout at Peterborough were evaluated in the same way as those brought about by road schemes, the one-year rate of return for 1978 would be some 57 per cent higher than the estimated commercial rate of return ([20] p. 160).

Of course, on many inter-urban routes there is competition between alternative public-transport modes, usually rail and bus or coach, and, on a few trunk routes, with air as well. Given the high costs of providing an air service, the cost of terminal capacity where this could be saved in the absence of domestic flights, and the externalities – of which noise and high fuel consumption are the most prominent – it is hard to believe that the benefits in terms of faster journeys for those passengers going by air are usually worth the costs, except where distances are well in excess of 300 miles. Indeed, purely in commercial terms, domestic air services and municipal airports typically make a loss [85]. The issue between rail and coach is more complicated. If all traffic were concentrated on rail, the extra costs might well be less than the savings in social cost from no longer operating a coach service. On the other hand, whilst rail is generally thought of as offering a superior service to the coach, there may be loss of benefits to passengers who find the coach more comfortable and convenient, and on some routes (especially cross-country routes) it may, in fact, be quicker. Were all traffic to be concentrated on the coach, the cost savings on rail may be much larger than this, particularly if the route is not to be retained for freight traffic, but in any case a major part of track maintenance costs is usually attributable to passenger services [86]. But if the traffic were heavy, costs, in terms of extra travelling time for passengers and – via increased congestion – other road users, may be large too. Which form of service is desirable will vary with the nature of the route, but it seems unlikely that both will be given the high overhead costs of providing a rail passenger service. The major exception may be where a rail passenger service is justified, but the coach caters for intermediate points not served by rail.

The important point to note is that no pricing system will necessarily lead to the elimination of one rival. Suppose that the services are very close substitutes and that both operators charge marginal cost. The traffic may be divided between them, although the extra benefits from having both services do not compensate for the extra overhead costs.[1] Or suppose both operators charge average cost. It may even be that both services will pay their way and yet the cost savings from eliminating one would exceed lost benefits. Thus there may be justification for using a system of licensing to protect the established operator from competition, such as exists in many countries including the

[1] Readers may recognise this as a typical example of the problem that increasing returns to scale pose for a competitive market system.

United Kingdom ([3] ch. 7). The obvious danger in this policy is that the monopolistic public-transport operator may become slack and give poor-quality service, or permit costs and fares to rise unreasonably. One role of the licensing authority is to prevent such slackness, but they have been widely criticised for failing to do this and, especially in the United States, for defending the interests of those they are designed to regulate [87]. Nor is there any guarantee that the licensing system will not protect the wrong mode, in terms of value for money. A further consequence of such licensing systems in terms of the distribution of real income must be borne in mind. Where it prevents the existence of road coach services at low fares, but rail continues to maximise the revenue from the service it operates, low-income families who would prefer a lower-quality, cheaper service will be harmed by the situation. Indeed, it may prevent them from undertaking long-distance journeys at all.

It must be said at this point that the discussion of inter-city travel so far is applicable particularly to the United Kingdom and may not be appropriate elsewhere. It is more difficult to divert passengers from car to public transport on leisure than on business journeys, since use of the car at the destination may be desired. The quality of feeder services will be relevant.[1] Most important of all, if potential traffic on the route in question is insufficient to justify provision of a fast and reasonably frequent train service, car or coach (depending on the income of the traveller) may remain dominant over much longer distances than in the United Kingdom.

THE RURAL TRANSPORT PROBLEM

The way in which the growth of private-passenger transport has precipitated what is commonly known as 'the rural transport problem' is by reducing public-transport usage to a level at which services are no longer commercially viable, and many have been (or are) threatened with being withdrawn. Figure 7 illustrates the pattern of decline.

[1] This raises most of the points considered in the previous chapter again, including the pattern of land-use and residential density.

In Figure 7 it is assumed that the demand is falling over time. The demand curve therefore shifts from position D_1 with fare P_1 to D_2 with the higher fare P_2 and finally to D_3. The spiral of costs and fares is similar to that for urban public transport, except that it has extended to the position where no combination of service frequency and fare can show a profit.

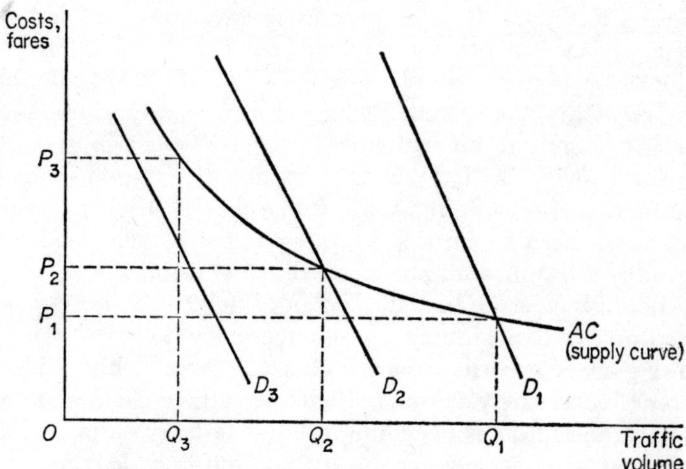

D_1 = initial demand curve

D_2 = shift in demand curve leads to higher fare and lower traffic flow

D_3 = further shift in demand curve leads to a situation where no fare increase will enable costs to be covered

FIG. 7 *Decline of rural transport*

At first sight the economist may tend to dismiss rural public transport as a product for which there is so little demand that it should not be produced at all. It is true that bus companies have typically followed average-cost-pricing policies even where they have much spare capacity, whilst rail fares have presumably been set to maximise revenue[1] (costs for a given rural network being assumed fixed). But even if public-transport fares and private-transport taxes were fully adjusted to marginal

[1] This is a somewhat doubtful statement, and has certainly only been true since the 1968 investigation by the Prices and Incomes Board [88]; prior to that, fares were at a fixed rate per mile regardless of route.

social cost, it is unlikely that much traffic in this sector would transfer between the two. The greater convenience of the car, particularly on journeys with shopping or heavy luggage, is too overwhelming for this to happen.

Even if the decline of rural public transport is justified on the usual economic resource-allocation criteria, it has very serious distributional implications. Those people in rural areas who cannot afford a car or who, because of age, non-availability of the family car[1] or other reasons do not drive, are suffering a major disbenefit in terms of loss of mobility. Moreover, mobility is not something for which other goods can readily compensate. Nor is it merely to be evaluated in terms of the willingness to pay for journeys actually made. Having the option of being able to make a shopping, visiting or other trip whenever one wants may itself be an important source of benefit [89]. Such a standby effect may also be enjoyed by car owners in case their car should happen to be out of action. Lack of public transport has probably contributed to the tendency for the younger working class to drift from villages to towns, whilst villages become increasingly populated with wealthier car commuters or with the buildings becoming used as holiday cottages [90]. Such tendencies are not necessarily bad, but they do completely change the character of rural villages, and as such have major non-market effects on their inhabitants.

Commentators on rural transport tend to stipulate the need for an 'adequate' service. Such a term is somewhat foreign to economists, who typically examine services in terms of their benefits and costs. The implication of the previous section is that such benefits are not well represented by fare revenue; provision of rural public transport should not be left to the test of profitability. Not that profitability has ever played a great role in determining the services to be provided. Major bus companies have long cross-subsidised rural services from urban and inter-urban ones. For instance, in the United Kingdom it was estimated that in 1958 the Tillings group of companies was losing £1·95 m. on rural services. One major operator alone, Midland Red, was losing money on 64 per cent of its route-mileage and

[1] Where the car is used for work journeys, wives and families may be left without transport for much of the working day [78].

32 per cent of bus miles run [91]. Indeed, it was a principle of the licensing authorities that bus companies would be granted monopolies on profitable routes in return for which they would be expected to 'use the fat to fry the lean'. To some extent, such figures for the degree of cross-subsidisation are misleading, since they are often based on a comparison of average revenue per bus mile on the route in question with average cost per bus mile for the company as a whole. Since costs vary greatly from route to route – in particular, the incremental costs of putting on off-peak rural services for shopping and visiting may be very low – a gross overstatement of costs may result.[1]

To the extent that cross-subsidisation does exist, and it was clearly government policy in 1968 that it should continue ([93] p. 21), it leads to higher fares on urban and inter-urban routes than would otherwise be necessary to break even, and may have contributed to the fact that public transport in towns with municipal companies has fared better than in similar towns without [60]. Similar cross-subsidisation existed on the railways, until in the wake of mounting deficits the Beeching plan of 1963 advocated major closures of rural services [94]. Whilst the costing of such services remained crude, and the exact level at which they may break even has been challenged [95], the general picture was clear. Even though some of the costs of the services were sunk, and could not be avoided by closure, simply the cost of maintenance of track and structure often outweighed fare revenue. More recently, passenger services on a number of rural lines have been subjected to cost–benefit analysis [96, 97]. Even allowing for the fact that some of the labour employed may be immobile, and that alternative employment may not be available in the area in question, so that the opportunity cost of employing it on rail services may be zero, the conclusion has not generally been favourable to the retention of rural passenger services, although the methods used, and in particular the omission of fare revenue as a benefit

[1] This situation has come to light in negotiations between bus companies and county councils; for instance, the Eastern Counties Omnibus Company originally asked for a subsidy of £47,000 per annum for a group of services in East Anglia, but after a re-examination of costs eventually agreed on £27,000 [92].

of operation in the Cambrian Coast study, have been the cause of much controversy [98, 99].

In the past decade, rural passenger transport has received considerable subsidisation. First and foremost, major grants have been made to British Rail, and on a smaller scale to bus companies, as direct subsidies. General measures to aid bus companies, such as fuel tax rebate and grants for new buses,[1] have also benefited rural users. One must doubt whether these grants have been wisely spent. Whilst the basis on which grants have been calculated for rail services is known typically to overstate the immediately avoidable cost of running a basic rural passenger service, the allocation of grants between rail and bus seems unduly biased towards the former. The result is that the historical accident of whether a town or village is on a secondary railway line (minor stations have largely been eliminated from main lines because stopping trains use track capacity that is wanted for express and freight trains) determines the quality of service and level of subsidy that it receives. The payment of a central subsidy for rail services, determined separately from the county's 'transport policy and programme', seems likely to further this misallocation of resources.

Another measure that has been advocated to aid rural public transport is the lifting of licensing restrictions on minibus operators. Since the cost savings from substituting a minibus for a full-size bus are not great – in particular, labour costs, which account for a large proportion of total costs, are little changed – and the minibus is often unsuitable for operating peak and non-rural services in between rural trips, their operation by major bus companies is seldom justified. But local operators combining their operation with private hire, garage and taxi work may be able to achieve greater flexibility and lower costs [92]. One encouraging recent development is the growth of post-bus services in the form of minibuses operated by the Post Office for the carriage of both mail and passengers.

It seems unlikely that anything short of drastic action in the form of regulation or motor-tax increases would have much effect on rural modal split, and in general such action would

[1] A prime aim of this grant was to encourage re-equipment with buses suitable for one-man operation [93].

not be justified. Although the decline of public transport is largely attributable to the growth of private transport, there is little case for suppressing the latter, solely as a way of preserving public transport. Direct subsidies, together with the searching-out of ways of providing adequate services at reduced cost, appear to be the best way of preserving such services as are deemed desirable. There is one major exception to this generalisation; this is the case of recreational trips to major tourist areas.

TRANSPORT AND RECREATION

The special characteristic of leisure trips to tourist areas is that they tend to be concentrated on specific routes and areas, and to be heavily peaked. For instance, on a peak day in 1965 Brighton attracted some 100,000 day visitors in addition to 20,000 longer-stay visitors; Rhyl and Prestatyn a peak of 110,000 day visitors and Southport 300,000 ([100] p. 94). The car has an advantage over public transport for the individual user not only in terms of convenience but also of cost on these trips; since leisure travel is mainly in family groups, car-occupancy rates are much higher than for other trips, and cost per person is proportionately less. Nevertheless, the external costs of congestion and pollution are large, whilst any new construction to cater for this traffic has to be justified in terms of benefits on only a few peak days each year. Reduced journey times resulting from motorway construction have already caused a major increase in day trips by car to the Lake District,[1] and is likely to have similar effects on Devon and the New Forest. Visitors to these areas do not necessarily wish to see congested village streets and country lanes with car-lined verges! Here is another classic example of how, in the presence of externalities, the results of the sum of individual optimal decisions may make everyone worse off. It would be totally ineffective for any individual to switch to public transport in order to relieve his destination of the effects of traffic; only

[1] Mansfield [101] estimates that the M6 caused a 30–40 per cent increase in trips. See also [102].

some form of collective action can do this. Moreover, the social costs of this type of traffic are not confined to spoiling the pleasure of other trippers at the time; particularly in areas such as the New Forest, which harbour many rare species of flora and fauna, motor traffic on this scale can do long-run or permanent damage to the ecology of the area. How to evaluate losses of this type is as yet largely an unanswered question in the economic appraisal of transport schemes. Many would argue that avoidance of such effects should be regarded as a constraint when planning for these areas.

As usual, possible solutions for this type of damage to the countryside may be grouped into pricing and administrative measures. Pricing would often be administratively feasible if still expensive; one would be charged a toll for entering an area such as the New Forest. The usual distributional problem, that this would reduce visits less the higher the income of the visitor, is reinforced in this case by a political notion that everyone has the 'right' to enjoy free access to such areas. As pedestrians usually create far less nuisance than motor vehicles, pedestrian charges would be much less – probably zero. Thus the use of public transport to such areas would be encouraged.

Alternatively, car visits to the area may be rationed or (administratively far cheaper) banned altogether. The economist's usual objection to outright banning is that it is unlikely that the optimal level of any externality is zero. But in this case, where the areas in question are valued partly for their 'wilderness' character in which *any* vehicles create intrusion, the social cost of admitting even a light flow of vehicles may be high. In any case the administrative cost of allowing a continued small flow may well not be justified. A compromise may be achieved by providing 'honey-pot' car parks with good facilities for motorists at selected points, and banning either parking or all traffic from other areas. Public transport may be provided if the area is large. For instance, the Goyt Valley scheme in the Peak District comprises free minibuses through the valley to the shore of a reservoir from a large car park [103]. Such a scheme, whilst protecting the area itself, does not relieve approach routes of heavy traffic in the way that a complete diversion of such trips to public transport would.

CONCLUSION

Whilst there are clear cases of price distortions leading to undesirable modal-choice decisions in the inter-urban passenger market as well as the urban, it is much less obvious how to deal with them. General lowering of rail fares and improvements in services may have as their major effect the redistribution of real income to the better-off sectors of the community, whilst elimination of so-called wasteful competition may lead to poorer services or higher fares. All the same, the separate planning of inter-urban road and rail facilities, using on the one hand cost-benefit and on the other commercial criteria, can hardly fail to distort resource allocation.

In the case of rural transport the growth of private transport may be responsible for the drastic fall in public-transport services, but curtailing private transport is not the best way of restoring them. Standard cost–benefit analysis has little to contribute in determining what is an adequate level of service; the value of economic analysis here is considering the most cost-effective way of providing such services as are deemed desirable and in devising methods of subsidisation that do not have unsought-for effects on decision-takers. The major exception is where an area is subject to heavy recreational use. In this case the argument for promoting a switch from private to public transport is stronger, and the possible measures to implement it are clearer.

7 Freight Transport

The frequently noted fact that road transport accounts for some 85 per cent of the tons of freight moved per annum in the United Kingdom may lead one to believe that the dominance of the lorry is so great that it is scarcely worth debating the issue of modal split in freight transport. This would be an unfortunate conclusion; in the first place, the dominance of the lorry is much reduced when length of haul is taken into account, lorries accounting for 61 per cent of freight ton-miles.[1] It must also be remembered that the actual choice of mode provides no evidence on the costs of diverting traffic to an alternative. For instance, it has been suggested that despite the dominance of railways in U.S. transport in 1890, the social cost of diverting all traffic to other modes would have been less than 5 per cent of G.N.P. [105]. If such a possibility existed now with respect to the lorry, many would regard the environmental benefits of the switch as being worth the social costs. So it is necessary to examine the issue more closely.

Unfortunately, analysis of the freight-transport sector is made extremely complicated by its diversity; whilst passenger journeys vary with respect to route, time and purpose, freight journeys also vary in the size, weight and form of the commodity carried. Although many commodities can be simply loaded into box vans, some require vehicles of a particular size or shape, or with special facilities such as refrigeration, liquid-carrying tanks with pumps, or tipping gear. It is to be expected, then, that – on these grounds alone – some traffic will be particularly suited to one mode of transport and some to another.

Traditionally, the key factor in explaining freight-transport modal split has been regarded as terminal costs. Road vehicles can be loaded at the point of origin of the trip and unloaded at

[1] Rail accounts for 18 per cent, coastal shipping for 16 per cent, and pipelines for 3 per cent ([104] p. 144).

the destination. Rail transport, on the other hand, requires either that the points of origination and destination be connected to the rail network, involving high costs of track installation and maintenance and of 'trip' working the traffic to and from the nearest main terminal, or that the freight be collected and delivered by road, with transhipment at railway depots. Provided that a route is already served by rail, and provides sufficient traffic to make up full train loads, the trunk part of the journey is probably cheaper by rail, but on shorter routes this advantage is lost in collection and delivery expenses, whilst repeated handling and shunting increases the risk of damage or delay.

The notion of a trade-off between terminal and line-haul costs determining the break-even length of haul for rail, for each type of traffic, is illustrated in Figure 8.[1]

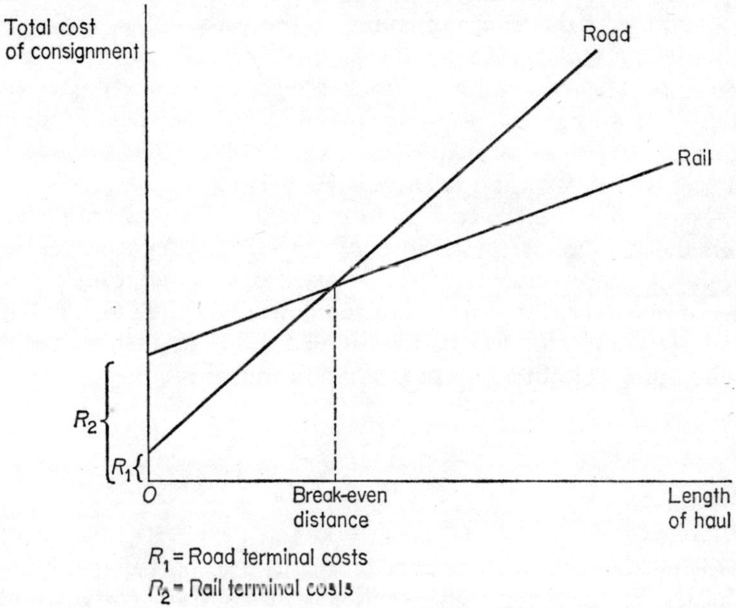

FIG. 8 *Freight costs by length of haul*

[1] This type of analysis was used recently in the New Zealand Transport Policy Study [106], where break-even length of haul varied from 50 miles for bulk liquid, 100 for logs, 150 for containers and up to 650 for livestock.

Obviously this is a gross simplification, but its implication for modal split is nevertheless appropriate. Rail transport will be more attractive the longer the route, the denser the traffic, and the easier the traffic is to load and unload. Where a particular traffic is sufficiently dense to make up into block trains, and the location of the origin and destination is such that these may readily be joined to the rail network, rail may be cheaper over quite short distances. Where transhipment is expensive it will only be on longer journeys that the cost savings on the trunk part of the journey make rail worthwhile.

Given this pattern of relative advantages, it is easy to see that the rapid growth of light-manufacturing industry at sites scattered around the peripheries of urban areas, and the relative decline of minerals and heavy industry, has favoured road transport rather than rail. For instance, by 1970 coal tonnages transported by rail in the United Kingdom were down to about two-thirds of the level prevailing in the early 1950s, although this traffic still accounted for 35 per cent of railway-freight receipts. The most common type of freight traffic now comprises short, dense movements; only about 8 per cent of all freight tonnage travels more than one hundred miles, whilst some two-thirds travels less than twenty-five ([104] ch. 3).

A broadly similar cost picture applies to water transport, whether by sea or canal. Here, of course, the geography of the route is more important still – many routes are difficult to serve by water transport – and the slowness of the mode makes it unattractive for freight that is perishable or of high value when the capital tied up in it while in transit is large.

PRICING POLICIES IN FREIGHT TRANSPORT

It is particularly with respect to freight transport that the unfairness of requiring railways to meet their track costs, whilst road hauliers had their track provided by the state, was first voiced. Moreover, to the extent that track costs do not vary with the level of traffic, loading them fully on to rail charges will lead to an inefficient allocation of traffic between modes

[107]. One drastic solution proposed was that the government should pay all railway track costs as well as those for roads [108]. However, unless the government controlled the size of the railway network, using the same cost–benefit analysis technique as it does for road, there would be a danger of inefficiency arising from the incentive this provides for the railways to maintain too much track capacity.

In any case the force of these arguments in the United Kingdom has been greatly reduced since the dismantling of the system of rail-freight regulation, which required railways to offer a standard rate for particular categories of traffic, regardless of the route and user concerned. At present, only some 2 per cent of railway-freight receipts, excluding parcels traffic, comes from traffic carried subject to national, published rates [109]. Prices for the remaining traffic are set by individual negotiation with the firms in question. In the past the principle has been that the charge should cover the direct costs of the traffic (fuel, labour and rolling stock) and make as large a contribution as possible to indirect costs (track, signalling and administration). Assuming it is designed to be a distinction between fixed and variable costs, this distinction is far from perfect [110]. The assumption is, presumably, that railways have spare track and signalling capacity which can be used at zero marginal cost, whilst all traffic requires extra rolling stock and labour. Clearly, in practice, there may be spare rolling stock and labour for part of the year, whilst track and signalling expenditure does vary with the density of traffic carried [86]. However, this 'pricing floor' was not, apparently, rigidly adhered to, and in present railways cost-accounting procedures a more flexible categorisation of costs according to their 'escapability' is followed ([104] ch. 3). Much traffic in recent years has been carried on a short-run basis that will not cover the full replacement cost of the rolling stock used.[1] Presumably this traffic will be given up when the current rolling stock ceases to be serviceable [109]. Such a system of price discrimination could, in theory, enable the railways to acquire all traffic which would accrue

[1] Indeed, the fact that in 1961 general merchandise revenue was estimated to fall £45·3 m. short of direct costs indicates that these floors had not been adhered to for some time ([94] p. 26).

to them, given a system of marginal cost-pricing, whilst yielding a higher revenue (Figure 9). In practice, the discrimination is undoubtedly less than perfect since the information of the railway authorities can never be perfect, whilst some customers are large enough to possess considerable bargaining power themselves. Thus revenue is lower than the maximum possible, and some traffic is lost that would have been willing to pay the marginal cost.

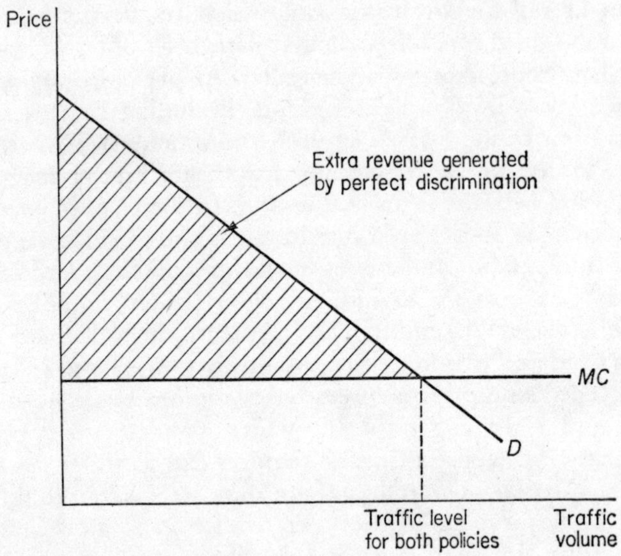

FIG. 9 *Price discrimination and marginal-cost pricing compared*

It has been argued that this pricing system is inappropriate, especially when applied to new traffic such as freight-liners [111], since it does not ensure that, even in the long run, all the costs of railway operation will be covered. However, as long as the indirect costs are to be incurred in any case, there is no more reason to expect new traffic to contribute to them than old. The issue hangs on an initial decision with respect to capacity. If it is correct to assume that the basic trunk network will remain in existence for the foreseeable future, then it is only incremental maintenance costs on the track that additional traffic should cover,[1] unless the network becomes congested or unless, in the absence of the traffic in question, capacity could be reduced.

The conclusion of this section is that, given a system of price discrimination operating on most traffic, there is no reason to suppose that rail charges are typically distorted in such a way as to lead to much traffic being lost for which the consignors' willingness to pay would more than cover the social costs of carrying it. In countries where the regulatory system prevents such discrimination this is not the case. In the road-freight-transport sector, most costs borne by the operators themselves are variable over a moderately short time period. Given the large numbers of firms operating in road haulage,[2] private and public, and the lack of any fixed freight rates, it seems likely that a similar pattern of traffic being carried, provided that it will cover its marginal private cost, will emerge there. So the real issue with respect to prices charged is again that of the relationship between road taxes and the externalities created by road transport.

REASONS FOR THE CHOICE OF FREIGHT-TRANSPORT MODE

As with passenger transport, the price paid is by no means the only determinant of the choice of mode. A major survey of the importance of various factors was carried out in 1966–7 [113]. The importance of quality of service is borne out by the discovery that many consignments travel by a mode known by the firm in question not to be the cheapest in terms of cost – usually by road haulier rather than by rail. Factors other than price frequently considered to be important are speed, reliability and freedom from damage or loss, all of which usually influence the choice towards road transport, though in this survey the latter

[1] Track costs of specific traffics arising from this cause should not be ignored as they may be substantial [86].

[2] In 1965 in the United Kingdom there were 46,442 public hauliers and 56,459 own-account hauliers with at least one vehicle over 3 tons unladen weight, although many were very small. One-half of the former and one-quarter of the latter possessed only one vehicle [112].

appeared insignificant. Special characteristics of the products needing specialised bodies influence the choice towards own-account road haulage, whilst suitability for containerisation makes public transport more likely.

However, the major determinants of the split between public and private transport appeared to be length of haul and consignment weight. The shorter the haul and the heavier the consignment the more likely the firm was to use its own road vehicles. Whilst the former relationship would be expected, the latter is perhaps surprising. The explanation appeared to be that very small consignments are usually sent by Post Office or passenger train, and it is only for heavier loads that it is worth using one's own vehicles.

THE DESIRABILITY OF A CHANGE IN MODAL SPLIT

Diesel lorries are major causes of the externalities discussed in Chapter 3, as well as having a greater effect on the speed of traffic than a private car (the equivalence rating usually given is one of three passenger-car units). Problems of noise and vibration are most acute with these vehicles, particularly in built-up areas. Whilst rail haulage produces some such effects, they are generally much less widespread. Thus there is probably considerable public support for the transfer of much traffic from road to rail.

As against this, a number of points may be made. It is scarcely conceivable that rail could perform urban collection and delivery work. Even before the rise of motor transport, rail had to be supplemented in this role by the horse and cart. But even on longer journeys, originating and terminating buildings are frequently not connected by rail. Rail terminals are usually sited towards the centres of the urban areas which have since grown up around them, whilst modern industrial estates are frequently on the peripheries of such areas. In these cases, diverting traffic to rail may, in fact, increase the mileage of urban road haulage needed, especially if the town in question is connected to the motorway network ([114] ch. 8).

The extent to which U.K. freight terminal facilities have been centralised, since the Beeching report advocated the closure of many smaller depots [94], undoubtedly cut the private costs of rail freight transport, but has increased the undesirable external effects. Consider the freight-liner network of express container trains. Terminals in London are situated at Stratford, Willesden, Kings Cross and Barking – generally well inside the built-up area. Moreover, individual depots only handle traffic for specific destinations, so that – for instance – traffic for the North-east from anywhere in the Greater London area has to be taken by road to Willesden. For a considerable number of locations it is likely that road haulage throughout would involve a reduced urban mileage in London.

Of course, this is not always the case. Where traffic is concentrated, terminals may be situated close at hand; for instance, the Southampton Docks freight-liner terminal is reached from the docks without traversing public roads. Traffic to and from private sidings does not use the road system at all. But the key point is that in the absence of special arrangements such as the current subsidy for the installation of private sidings, railways – like other operators – are given no incentive to reduce the social, as opposed to the private, cost of their operations. The 1962 Road Goods Survey found some 10,170 m. ton-miles of road haulage on end-to-end trips of over 100 miles [115]. It seems likely that even if there are good social reasons for some of this traffic to use road transport, there exists a considerable (and growing) amount of traffic basically suitable for rail transport for which the external costs of carriage could be reduced if suitable collection and delivery arrangements could be made.[1] It is necessary to point out that more than half of this is already carried by public transport, in the sense of firms offering a transport service to other firms rather than carrying their own freight. Use of public hauliers may offer environmental advantages over private transport if higher load factors and less-empty running result from the increased ability to get a return load. But on the whole, it is with diversion from road

[1] For the railways' own estimates of traffic flows by road that were suitable for rail carriage at this time see [116].

to another mode of transport that we are concerned with in this chapter.

Little has been said here so far with respect to water transport, despite the importance of coastal shipping in the U.K. freight market, and the environmental attractions of both this and canals. Much that has been said concerning railway transport applies here too. But the main limitation on the usefulness of these forms of transport is their inability to serve many routes, at any rate without major investment in new or reconstructed canals. In some cases, such investments may be justified in terms of environmental benefits if not on commercial grounds,[1] but frequently their major competitors are railways, over which their environmental advantages are less great.

POLICIES TO ALTER THE MODAL SPLIT OF FREIGHT TRANSPORT

We have suggested that the freight-transport market offers a typical picture of how the presence of external costs leads to a sub-optimal allocation of traffic between modes, as well as to decisions on freight-handling practices which may fail to minimise the social cost of the journey in question. Since it is the response of organisations to current market prices that leads to these outcomes, the simplest way of dealing with the problem may again seem to be by the imposition of taxes to equate social and private cost.

Unfortunately, all the problems with taxation discussed above (Chapter 4) apply with equal or even greater force in this sector. There are already substantial taxes on road-freight transport operators in the form of fuel tax and vehicle-licence duty. Payments of the former are obviously related in some way to vehicle use and size, and the latter varies by unladen vehicle weight (at present from £24 up to £459 per annum). It is often

[1] A major scheme that has been put forward recently is that of enlarging the Sheffield and Yorkshire Navigation between Doncaster and Rotherham [117]. Another idea put forward, mainly on environmental grounds, is the use of the London canal system to transport freight from the M1 motorway at Watford to London Docks.

argued that, at least for some traffic, these more than cover the external costs of journeys made, and thus the tax system biases modal choice towards the railways. The big difficulty is that neither of these taxes can be varied according to the nature of the roads on which the vehicle is running. A full road pricing system for freight vehicles, even in a limited area, would require many more vehicles to be equipped with meters or extra licences than is the case with the private car, where local car commuters may form the crux of the problem. Solutions such as increased parking charges or controls are unlikely to help, as the relevant journeys it is desired to suppress are mainly through-journeys.

One argument put forward in favour of the system of licensing common throughout the world and in existence in the United Kingdom until the passage of the 1968 Transport Act was that, by requiring evidence of need for the service before a road haulier would be licensed to set up in business or take on additional vehicles, it prevented road transport from carrying traffic for which the railways offered a suitable service. This argument was rejected by the Geddes Committee [118] on the grounds that the licensing system freely permitted the use of own-account road transport. Moreover, by preventing own-account operators from seeking part loads and return loads from other firms, the licensing system could increase the amount of empty running of lorries and thus cause additional congestion and pollution. In fact, its elimination appears to have led to little carriage for other firms on the part of own-account operators [119].

Since the 1968 Act was implemented, licences have been freely available, subject to proof of adequate financial resources and vehicle-maintenance facilities. It was originally proposed that an additional 'quantity' licence would be required for any vehicle of more than sixteen-tons gross weight carrying goods a distance of more than one-hundred miles, or for certain goods – coal, iron ore, steel, and so forth – for any distance. This would apply to own-account operators as well as public-haulier operation. These licences would only be issued if it were found that road haulage could not be matched by rail in terms of price or quality of service. The interesting point about this system is therefore that it was not apparently envisaged as a way of taking into account external costs in the allocation of traffic between modes. Traffic would only be diverted where it was in the firm's own interests to do so [11]. The need for such a system would only arise if firms were, because of ignorance or prejudice, failing to optimise in the use of transport. There is evidence that firms are often unaware of the costs of operating their own vehicles and of the characteristics of the alternatives available, although most own-account operators make some use of public hauliers and/or rail transport, and are not totally ignorant of their potential [113, 120]. But it is doubtful whether such a licensing system, given all the complications of weighing up intangibles with respect to the quality of service, would in the end transfer a large amount of traffic to rail.

In fact the quantity licensing system contained in the 1968 Act has never been utilised. But some such system could be implemented with instructions to take into account external costs as well. The administrative cost would be high, since a detailed case-by-case approach would be needed; firms would have to provide cost information on a scale that would be difficult, especially for own-account operators [121], and the familiar problem of measurement and valuation of environmental costs would again emerge. If there was to be any coherence to the decisions of the licensing authorities, methods for the latter would have to be laid down by the Department of the Environment. But despite all its costs such a system may still be worthwhile.

The other measure of co-ordination in the 1968 Act was the merging of the road collection and delivery services and the freight-liner company of the railways with the nationalised road-haulage companies to form the National Freight Corporation. The aim was to yield considerable cost savings through the elimination of duplication of services; but, although the resulting company was ordered to use rail 'wherever economically justified' [122], it is unlikely that this will lead to the transfer of much traffic from road to rail, and it may have the reverse effect. It is still not the case that external costs are to be taken into account in the allocation of traffic, and whilst the existence of a large nationalised freight company must make doing so easier, to impose upon it a vague social obligation without clear guidance and financial control would be to invite similar problems to those that emerged with the British Transport Commission in the 1950s [123].

A different method of influencing modal split was adopted in 1974 when the U.K. government consulted with major companies to discover the scope for switching traffic from road to rail. Given the dominance of large firms amongst freight transport users,[1] this would seem to be a much simpler way of bringing publicity to the potential of rail services and applying suasion for their use. But it still does not cope with the divergence of private and social costs, except to the extent that subsidies are to be provided for the installation of new private sidings. It is unlikely that many firms can be persuaded to use services that are more expensive or lower in over-all quality to themselves without compensation.

OTHER MEASURES TO MITIGATE THE EXTERNAL EFFECTS OF ROAD FREIGHT TRAFFIC

The other measures available to mitigate the external effects of road freight transport may again be grouped into changes to vehicle design and changes to the road network. Apart from measures to reduce the noise and air pollution from freight vehicles, other possibilities are to reduce congestion by increasing the minimum permitted power–weight ratio from its current level of 6 brake horse power per ton,[2] or to alter the legally permitted maximum gross vehicle weight from its current level of thirty-two tons. Raising this may lead to a smaller number of vehicles with an increased social cost per vehicle, lowering it to the reverse. A further option is to alter the legally permitted weights for specific roads or areas. At present such measures are confined to banning through movements of heavy lorries in certain cities, but there are plans to

[1] At present, one-half of railway-freight revenue comes from thirty customers and three-quarters from 300 ([109] p. 235).

[2] A study of this found that such a further change would not be worthwhile [124].

designate specific lorry routes which they must use except when making a delivery or collection off the network. To the extent that any of these measures changed the private costs of road transport, they would also have repercussions for modal split.

A more drastic measure would be to bar lorries above a certain weight from specific urban areas. This would probably be associated with the setting up of peripheral transhipment depots, where loads could be transferred to small (possibly electrically propelled) vehicles. It is sometimes argued that this would replace a small number of large vehicles with many small ones. However, large vehicles often enter urban areas in which only a small part of their load is to be delivered. Moreover, loads could be rearranged at the transhipment depot so that whole loads for each destination could be built up for delivery in one block outside peak hours instead of a number of separate lorries calling. An experiment along these lines is being conducted by National Carriers Limited in Oxford ([125] p. 60).

The private costs of such changes may be large, but it is necessary to create an institutional structure in which they can be weighed against the environmental benefits. Should they be found worthwhile, by making transhipment necessary for the use of bulk road transport for trunk journeys, one disadvantage of rail in this sector would be removed. If the transhipment depots were served by private sidings or located close to rail terminals, it is possible that rail might become attractive for trunk haulage. Similarly, locating new industrial complexes at points with ready rail access may yield external benefits.

The other way of attempting to remove heavy vehicles from areas in which they inflict external costs on others is, of course, to build new roads. We have indicated above some of the reasons why this may be both socially expensive and of limited effectiveness. Within urban areas, for instance, bypassing shopping centres and providing rear access for loading and unloading may provide benefits to shoppers [126]. But without substantial redevelopment it is not usually possible to provide routes that are thoroughly segregated from residents and workers in urban areas. Such redevelopment has its own external costs (see Chapter 3).

CONCLUSION

The argument of this chapter has not been that there is any simple solution to the problem of the transport of freight, such as the wholesale banning of the use of road transport. For a large proportion of traffic there is no alternative, and railways themselves make widespread use of road vehicles for collection and delivery work. What we have argued is that there may well be a considerable volume of traffic travelling by road which would best be transferred to rail if its full social cost could be calculated. Moreover, existing patterns of prices lead both road transport and railways to use freight-handling methods and locations that may not minimise social costs. In the absence of a method of reflecting the full social costs in prices paid by freight-transport users, there is a need to consider, on an individual-area and individual-traffic basis, the social costs and benefits of changes in the pattern of freight transport, and to make institutional arrangements for desirable changes to be implemented.

Although such changes as have been considered here may increase the money costs of transport, and thus raise the retail prices of goods, there would be considerably offsetting benefits. It is calculated that, in 1965, freight transport as a whole accounted for 6·5 per cent of value added in the U.K. economy, and road freight transport for 5·5 per cent ([112] p. 3). Of course, for manufacturing the average is higher – some 9 per cent – and varies greatly, from 1 per cent in the case of clocks and watches to 50 per cent for some quarrying activities [127, 128]. But it is clear that even a large proportionate increase in the money costs of transport would not have the huge effect on prices that is sometimes feared.

8 Conclusion

The theme of this book has been that transport decisions taken on the single criterion of profitability will be seriously deficient. The reasons for this are (1) that the transport sector is subject to major indivisibilities, joint costs and economies of scale which mean that marginal private cost may often be well below average, and (2) that transport systems are major creators of externalities, which are not taken into account by individuals in their market behaviour.

If it were possible to reflect fully all these discrepancies between market prices and marginal social cost by means of taxes, then one could leave transport decisions to the individual user acting on these signals. Incentives would be given for fitting anti-pollution devices, switching to electric traction, using transhipment depots, or transferring from private to public transport wherever such measures were desirable. However, the administrative costs of levying such a comprehensive system of taxes, which would have to vary with type of vehicle, location of use, time of day of use, and so forth, would almost certainly prevent it from ever being introduced. Whilst some form of road pricing might be worthwhile, it would almost certainly be a much more limited version involving considerable averaging and being confined to specific urban areas.

If this is the case then the door is opened to a wide variety of other measures, including second-best pricing policies for public transport – where administration of complicated fares structures is somewhat cheaper – legislative and administrative controls, and licensing systems. The aim of these would be partly to transfer traffic from private to public transport, but also to alter the over-all demand for transport facilities and the techniques used in running them.

Such changes will impose costs on members of society as

consumers, in the form of slower or less-convenient journeys and higher prices for some goods. On the other hand, environmental benefits will accrue, often to the same persons. The point is that, at present, they typically have no way of registering their preferences in terms of the environment, or of showing what material sacrifices they would be willing to make to improve it.[1] The only markets in which such preferences can be revealed at all are those concerned with geographical location – that is housing and employment – which are themselves very imperfect markets and ones in which all consumers operate only subject to very severe constraints.[2]

The first requirement from the point of view of implementing such policies is clearly to create an institutional structure in which there are authorities with the power and responsibility to examine and introduce them. In the past provision of public transport – whether privately or state-owned – has typically been controlled by a completely separate organisation from that responsible for road-building and traffic management, whilst licensing authorities have been distinct again. Such subsidies as have been available may well have not been put to the best use because they have been earmarked for specific railway lines or capital expenditure. The requirement for local authorities to produce comprehensive 'transport policies and programmes' for which they will be allocated block grants is therefore a very encouraging development, although most rail subsidies are still excluded from this system.

Of course, control of transport by the national government and/or local authorities on the scale suggested here has its dangers. There is no necessary guarantee that their decisions will be soundly based. Despite the growing use of cost–benefit analysis, the state of the art in transport decision-taking still

[1] An interesting approach to research in this area is provided by the 'priority evaluator technique', which consists of placing participators in a hypothetical situation in which they can 'buy' improvements in transport or in their environment. Early results show that many people would be willing to suffer poorer transport facilities in order to improve their environment [28].

[2] House prices vary not only with the characteristics of the house but also with environmental factors such as the density of population in the area, noise, air pollution and proximity to a green belt [129, 130].

leaves a lot to be desired [46]. Direct public participation in transport planning, and in the supposed ultimate safeguard, the public enquiry, is still heavily dominated by pressure groups and the more articulate members of society. But the diverse nature of the transport market and of the geographical layout of towns means that there is no real alternative to careful detailed planning of transport systems for specific areas. No simple, nationwide measure to transfer traffic from private to public transport is likely to be justified; but in plans for individual areas and individual traffics, it is the author's view, admittedly based on inadequate evidence, that a significant transfer of traffic on these lines should take place.

Further Reading

There are many good introductions to transport economics in general, including Thomson [67] and Heggie [25]; for more emphasis on the historical and institutional, as well as the purely economic issues, see Gwilliam and Mackie [3]. A book which concentrates on methods of evaluation and the political and planning framework within which they are used is A. J. Harrison, *The Economics of Transport Appraisal* (London: Croom-Helm, 1974). The classic U.K. official publication on the track-costs debate is [10], but for a thorough study of the economics of road-user charges, see Walters [15]. On the external costs of transport systems see Lassiere and Bowers [28]. Two O.E.C.D. publications, [26, 42], deal with pollution and resource depletion respectively. For a consideration of the consequences of transport policy for the majority who do not have a car available, see Hillman [78]. A wide-ranging collection of papers on urban passenger and freight transport is contained in *Urban Transport*, ed. D. A. Hensher (Cambridge University Press, 1976). Finally, the soundest non-technical statement of the case for diverting traffic from private to public transport is contained in the Report of the Independent Commission on Transport [20].

Bibliography

[1] E. J. Mishan, *The Costs of Economic Growth* (Harmondsworth: Penguin, 1967).

[2] G. Roth, *Paying for Roads; the Economics of Traffic Congestion* (Harmondsworth: Penguin, 1967).

[3] K. M. Gwilliam and P. J. Mackie, *Economics and Transport Policy*, (London: Allen & Unwin, 1975).

[4] Royal Commission on Transport, *Final Report*, Cmd. 3751 (London: H.M.S.O., 1930).

[5] M. Webb, *Transporting Goods by Road* (London: Weidenfeld & Nicolson, 1972).

[6] C. A. Nash, D. W. Pearce and J. K. Stanley, 'An Evaluation of Cost–Benefit Analysis Criteria', *Scottish Journal of Political Economy*, vol. 22 (June 1975).

[7] D. W. Pearce, *Cost-Benefit Analysis* (London: Macmillan, 1971).

[8] C. H. Sharp, *Transport Economics* (London: Macmillan, 1973).

[9] British Railways Board, *A Study of the Relative True Costs of Rail and Road Freight Transport over Trunk Routes* (London, 1964).

[10] Ministry of Transport, *Road Track Costs* (London: H.M.S.O., 1964).

[11] D. L. Munby, 'Mrs. Castle's Transport Policy', *Journal of Transport Economics and Policy*, vol. 2 (1968).

[12] A. A. Walters, 'The Long and Short of Transport', *Bulletin of the Oxford University Institute of Economics and Statistics*, vol. 27 (1965).

[13] N. Lee and M. Dalvi, 'Variations in the Value of Travel Time', *Manchester School*, vol. 37 (1969).

[14] A. A. Walters, 'The Theory and Measurement of Private and Social Cost of Highway Congestion', *Econometrica*, vol. 29 (1961).

[15] A. A. Walters, *The Economics of Road User Charges*, World Bank Staff Occasional Papers No. 5 (Baltimore: Johns Hopkins Press, 1968).

[16] J. C. Tanner, *An Economic Comparison of Motorways with 2 or 3 Lanes in Each Direction*, Road Research Laboratory No. LR 203 (London: H.M.S.O., 1968).

[17] O. E. Williamson, 'Peak-Load Pricing and Optimal Capacity under Indivisibility Constraints', *American Economic Review*, vol. 56 (1966).

[18] J. R. Meyer, M. J. Peck, J. Stenason and C. Zwick, *The Economics of Competition in the Transport Industries* (Harvard University Press, 1959).

[19] P. Cottinet, 'The Allais Report', *Journal of Transport Economics and Policy*, vol. 1 (1967).

[20] Independent Commission on Transport, *Changing Directions: A Report* (London: Coronet Books/Hodder Paperbacks, 1974).

[21] P. A. Victor, *Economics of Pollution* (London: Macmillan, 1972).

[22] D. N. M. Starkie and D. M. Johnson, *The Economic Value of Peace and Quiet* (London: Saxon House, 1975).

[23] J. Diffey, *An Investigation into the Effect of High Traffic Noise on House Prices in a Homogeneous Sub-Market*, Centre for Environmental Studies (1971).

[24] Sir Alan Wilson, *Noise: Final Report of the Committee on the Problem of Noise*, Cmnd. 2056 (London: H.M.S.O., 1963).

[25] I. G. Heggie, *Transport Engineering Economics* (New York: McGraw-Hill, 1972).

[26] O.E.C.D., *Effects of Traffic and Roads on the Environment in Urban Areas* (Paris, 1973).

[27] R. Ridker, *The Economic Costs of Air Pollution* (New York: Praeger, 1967).

[28] A. Lassiere and P. Bowers, *Studies on the Social Costs of Urban Road Transport*, Report of the Eighteenth Round Table on Transport Economics, European Conference of Ministers of Transport (1972).

[29] A. J. Harrison and D. A. Quarmby, 'The Value of Time', in *Cost–Benefit Analysis*, ed. R. Layard (Harmondsworth: Penguin, 1972).

[30] R. F. F. Dawson, *Current Costs of Road Accidents in Great Britain*, Road Research Laboratory Report No. LR 396 (Crowthorne: Department of the Environment, 1971).
[31] J. R. Meyer, J. F. Kain and M. Wohl, *The Urban Transportation Problem* (Harvard University Press, 1965).
[32] Jane Jacobs, *The Death and Life of Great American Cities* (Harmondsworth: Penguin, 1962).
[33] P. A. Victor, *Pollution: Economy and Environment* (London: Allen & Unwin, 1972).
[34] A. Please and D. C. Pike, *The Demand for Road Aggregates*, Road Research Laboratory Report No. LR 185 (London: H.M.S.O., 1968).
[35] M. Whitbread and H. Bird, 'Rent, Surplus and the Evaluation of Residential Environments', *Regional Studies*, vol. 7 (1973).
[36] P. Davidson, 'Public Policy Problems of the Domestic Crude Oil Industry', *American Economic Review*, vol. 53 (1963).
[37] R. Nield, 'Replacement Policy', *National Institute Economic Review*, no. 30 (1964).
[38] C. A. Nash, 'Future Generations and the Social Rate of Discount', *Environment and Planning*, vol. 5 (1973).
[39] R. G. Harman, 'Fuel in Transport', *Traffic Engineering and Control*. vol. 15 (1974).
[40] W. P. Goss and J. C. McGowan', 'Transportation and Energy – A Future Confrontation, *Transportation*, vol. 1 (1972).
[41] I. C. Cheeseman, 'Transport Technology – Master or Servant', *Chartered Institute of Transport Journal*, vol. 35 (1973).
[42] G. Leach, *The Motor Car and Natural Resources* (O.E.C.D., 1972).
[43] H. S. D. Cole *et al.*, *Thinking about the Future: A Critique of the Limits to Growth* (London: Chatto & Windus, 1973).
[44] G. Hoinville, 'Evaluating Community Preferences', *Environment and Planning*, vol. 3 (1971).
[45] P. Healy, 'Social Impact Analysis in the Planning Process', in *Urban Transport*, ed. D. A. Hensher (Cambridge University Press, 1976).

[46] J. K. Stanley and C. A. Nash, 'The Evaluation of Transport Improvements', in *Urban Transport*, ed. D. Hensher (Cambridge University Press, 1976).
[47] I. Hoch, 'Inter-Urban Differences in the Quality of Life', in *Transport and the Urban Environment*, ed. J. Rothenburg and I. G. Heggie (London: Macmillan, 1974).
[48] H. W. Richardson, 'A Note on the Distributional Effects of Road Pricing', *Journal of Transport Economics and Policy*, vol. 8 (1974).
[49] W. J. Tyson, 'The Peak in Road Passenger Transport', *Journal of Transport Economics and Policy*, vol. 6 (1972).
[50] Ministry of Transport, *Road Pricing – the Economic and Technical Possibilities*, Smeed Report (London: H.M.S.O., 1964).
[51] J. M. Thomson, 'An Evaluation of Two Proposals for Traffic Restraint in Central London', *Journal of the Royal Statistical Society*, series A, vol. 130 (1967).
[52] R. Turvey, *Economic Analysis and Public Enterprises* (London: Unwin, 1971).
[53] C. D. Buchanan, *Traffic in Towns* (London: H.M.S.O., 1963).
[54] M. E. Beesley and J. F. Kain, 'Urban Form, Car Ownership and Public Policy: An Appraisal of Traffic in Towns', *Urban Studies*, vol. 1 (1964).
[55] J. M. Thomson, 'The Value of Traffic Management', *Journal of Transport Economics and Policy*, vol. 2 (1968).
[56] D. A. Hensher, 'The Demand for Ubran Passenger Transport', in *Urban Transport*, ed. D. Hensher (Cambridge University Press, 1976).
[57] E. Smith, 'Economics of Urban Railways and Express Buses', *Journal of Transport Economics and Policy*, vol. 7 (1973).
[58] Economic Planning Directorate, *The Value of Time Savings in Transport Investment Appraisal*, Technical Note 3 (Ministry of Transport, 1969).
[59] D. A. Hensher, 'The Consumer's Choice Function: A Study of Traveller Behaviour and Values', unpublished Ph.D. thesis (University of New South Wales, 1972).

[60] P. R. White, 'Use of Public Transport in Britain & Ireland', *Journal of Transport Economics and Policy*, vol. 8 (1974).

[61] F. V. Webster, *A Theoretical Estimate of the Effect of London Car Commuters Transferring to Bus Travel*, Road Research Laboratory Report No. LR 165 (1968).

[62] C. D. Foster and M. E. Beesley, 'Estimating the Social Benefit of Constructing an Underground Railway in London', *Journal of the Royal Statistical Society*, series A, vol. 126 (1963).

[63] Colin Buchanan and Partners/Freeman, Fox, Wilbur Smith and Associates, *Alternatives for Edinburgh* (Edinburgh, 1971).

[64] N. Lichfield and Honor Chapman, 'Urban Transport and Modal Choice', *Journal of Transport Economics and Policy*, vol. 7 (1973).

[65] D. W. Pearce and C. A. Nash, 'The Evaluation of Urban Motorway Schemes: A Case Study – Southampton', *Urban Studies*, vol. 10 (1973).

[66] A. Downs, 'Uncompensated Non-Construction Costs which Urban Highways and Urban Renewal Impose upon Residential Households', in *Analysis of Public Output*, ed. J. Margolis (New York: National Bureau of Economic Research, 1970).

[67] J. M. Thomson, *Modern Transport Economics* (Harmondsworth: Penguin, 1975).

[68] C. D. Foster and P. J. Mackie, 'Noise: Economic Aspects of Choice', *Urban Studies*, vol. 7 (1970).

[69] J. S. Bain, 'The Technology, Economics and Industrial Strategy of Automotive Air Pollution Control', *Western Economic Journal*, vol. 8 (1970), reprinted in J. S. Bain, *Environmental Decay* (Boston: Little, Brown, 1973).

[70] H. J. Baum, 'Free Public Transport', *Journal of Transport Economics and Policy*, vol. 7 (1973).

[71] L. N. Moses and H. F. Williamson, 'Value of Time, Choice of Mode and the Subsidy Issue in Urban Transportation', *Journal of Political Economy*, vol. 71 (1963).

[72] G. Kraft and T. A. Domenciçh, 'Free Transit', in

Readings in Urban Economics, ed. M. Edel and J. Rothenburg (New York: Macmillan, 1972).

[73] M. A. Cundill and P. F. Watts, *Bus Boarding and Alighting Times*, Transport and Road Research Laboratory Report No. LR 521 (1973).

[74] P. R. White, 'Economics of Urban Railways and Express Buses – a Comment', *Journal of Transport Economics and Policy*, vol. 7 (1973) with rejoinder by E. Smith in same issue.

[75] S. Trench and J. A. Slack, 'Nottingham's New Transport Policy', *Traffic Engineering and Control*, vol. 95 (1973).

[76] H. J. Wootton and G. W. Pick, 'A Model for Trips Generated by Households', *Journal of Transport Economics and Policy*, vol. 1 (1967).

[77] C. A. Nash, 'The Treatment of Capital Costs of Vehicles in Evaluating Road Schemes', *Transportation*, vol. 3 (1974).

[78] M. Hillman with I. Henderson and A. Whalley, *Personal Mobility and Transport Policy*, Political and Economic Planning Broadsheet 542 (1973).

[79] M. E. Beesley and C. D. Foster, 'The Victoria Line; Social Benefit and Finances', *Journal of the Royal Statistical Society*, series A, vol. 128 (1965).

[80] *Second Report of the Expenditure Committee of the House of Commons Session 1972–3, on Urban Transport Planning* (London: H.M.S.O., 1973).

[81] A. Evans, 'Inter-City Travel and the London Midland Electrification', *Journal of Transport Economics and Policy*, vol. 3 (1969).

[82] P. A. Stanley and P. R. White, *A Review of Medium Density Inter-Urban Transport*, Transport Studies Group, Polytechnic of Central London (1973).

[83] G. A. Searle, 'COBA, A Computer Program for the Economic Assessment of Road Schemes', *Traffic Engineering and Control*, vol. 13 (1972).

[84] D. W. Glassborow, 'The Road Research Laboratory's Investment Criteria Examined', *Bulletin of the Oxford University Institute of Economics and Statistics*, vol. 22 (1960).

[85] R. S. Doganis and G. F. Thompson, *The Economics of British Airports*, Transport Studies Group, Polytechnic of Central London (1973).

[86] S. Joy, 'British Railways' Track Costs', *Journal of Industrial Economics*, vol. 13 (1964).

[87] D. F. Pegrum, *Transportation ; Economics and Public Policy*, 3rd edn (Homewood, Ill.: Irwin, 1973).

[88] National Board for Prices and Incomes, *British Railways Fares and Charges*, Report No. 72 (London: H.M.S.O., 1968).

[89] B. Weisbrod, 'Collective Consumption Services of Individual Consumption Goods', *Quarterly Journal of Economics*, vol. 28 (1964).

[90] D. St John Thomas, *The Rural Transport Problem* (London: Routledge & Kegan Paul, 1963).

[91] Ministry of Transport, *Rural Bus Services* (London: H.M.S.O., 1961).

[92] J. Hibbs, 'Maintaining Transport Services in Rural Areas', *Journal of Transport Economics and Policy*, vol. 6 (1972).

[93] *Public Transport and Traffic*, Cmnd. 3481 (London: H.M.S.O., 1967).

[94] British Railways Board, *The Reshaping of British Railways* (London: H.M.S.O., 1963).

[95] D. L. Munby, 'The Reshaping of British Railways', *Journal of Industrial Economics*, vol. 11 (1963).

[96] Ministry of Transport, *The Cambrian Coast Line* (London: H.M.S.O., 1969).

[97] P. K. Else and M. Howe, 'Cost–Benefit Analysis and the Withdrawal of Railway Services', *Journal of Transport Economics and Policy*, vol. 3 (1969).

[98] R. Sugden, 'Cost–Benefit Analysis and the Withdrawal of Railway Services', *Bulletin of Economic Research*, vol. 24 (1972).

[99] K. Richards, 'The Economics of the Cambrian Coast Line', *Journal of Transport Economics and Policy*, vol. 6 (1972).

[100] J. Allan Patmore, *Land and Leisure* (Harmondsworth: Penguin, 1972).

[101] N. W. Mansfield, 'Recreational Trip Generation', *Journal of Transport Economics and Policy*, vol. 3 (1969).
[102] I. R. Gordon and S. L. Edwards, 'Holiday Trip Generation', *Journal of Transport Economics and Policy*, vol. 7 (1973).
[103] Countryside Commission, *Goyt Valley Traffic Experiment* (London: H.M.S.O., 1970).
[104] A. W. J. Thomson and L. C. Hunter, *The Nationalised Transport Industries* (London: Heinemann, 1973).
[105] R. W. Fogel, *Railroads and American Economic Growth: Essays in Econometric History* (Baltimore: Johns Hopkins Press, 1964).
[106] Wilbur Smith and Associates, *New Zealand Transport Policy Study*, Ministry of Transport, New Zealand (1973).
[107] W. J. Baumol *et al.*, 'The Role of Cost in the Minimum Pricing of Railway Services', *Journal of Business*, vol. 35 (1962).
[108] H. O. Mance, 'The Road and Rail Transport Problem', *Journal of the Institute of Transport*, vol. 21 (1941).
[109] S. Joy, 'Pricing and Investment in Railway Freight Services', *Journal of Transport Economics and Policy*, vol. 5 (1971).
[110] C. D. Foster, 'Some Notes on Railways Costs and Costing', *Bulletin of the Oxford University Institute of Economics and Statistics*, vol. 24 (1962).
[111] A. J. Harrison, 'Investment in Liner Trains', *Bulletin of the Oxford University Institute of Economics and Statistics*, vol. 26 (1964).
[112] S. L. Edwards and B. T. Bayliss, *Operating Costs in Freight Transport* (Department of the Environment, 1972).
[113] S. L. Edwards and B. T. Bayliss, *Industrial Demand for Transport*, Ministry of Transport (London: H.M.S.O., 1970).
[114] C. Sharp, *Living with the Lorry*, Transport Research Unit (Leicester University, 1973).
[115] Ministry of Transport, *Survey of Road Goods Transport, 1962; Final Results Part 1* (London: H.M.S.O., 1964).
[116] British Railways Board, *The Development of the Major Railway Trunk Routes* (London, 1965).

[117] T. E. Phillips, 'Freight on Britain's Internal Waterways', *Chartered Institute of Transport Journal*, vol. 35 (1973).
[118] Ministry of Transport, *Report of the Committee on Carriers' Licensing* (London: H.M.S.O., 1965).
[119] B. T. Bayliss, *The Road Haulage Industry Since 1968*, Department of the Environment (London: H.M.S.O., 1973).
[120] C. Sharp, *The Allocation of Freight Traffic – A Survey*, Ministry of Transport (1970).
[121] G. W. Fernando, 'The Use of Own-Account Operators' Accounting Information for Quantity Licensing', unpubl. M.Phil. thesis (Leeds University, 1971).
[122] *The Transport of Freight*, Cmnd. 3470 (London: H.M.S.O., 1967).
[123] S. Joy, *The Train That Ran Away* (London: Ian Allan, 1973).
[124] C. Sharp and A. Jennings, 'More Powerful Engines for Lorries?', *Journal of Transport Economics and Policy*, vol. 6 (1972).
[125] Department of the Environment, *Lorries and the World We Live In* (London: H.M.S.O., 1973).
[126] A. Jennings, C. Sharp and D. Whitby, *Delivering the Goods: A Study of Watford Service-Only Precinct*, Freight Transport Association Research Paper No. 2 (1971).
[127] S. L. Edwards, 'Transport Costs in British Industry', *Journal of Transport Economics and Policy*, vol. 4 (1970).
[128] S. L. Edwards, 'Transport Costs in the Wholesale Trades', *Journal of Transport Economics and Policy*, vol. 3 (1969).
[129] D. W. Pearce, 'The Economic Evaluation of Noise-Generating and Noise-Abatement Projects', in *Problems of Environmental Economics* (Paris: O.E.C.D., 1972).
[130] J. S. Wabe, 'A Study of House Prices as a Means of Establishing the Value of Journey Time, the Rate of Time Preference and the Valuation of Some Aspects of the Environment in the London Metropolitan Region', *Applied Economics*, vol. 3 (1971).